MEDICAL ETHICS IN ANTIQUITY

A PALLAS PAPERBACK

PAUL CARRICK

Department of Philosophy, Temple University, Philadelphia

MEDICAL ETHICS IN ANTIQUITY

Philosophical Perspectives on Abortion and Euthanasia

D. REIDEL PUBLISHING COMPANY

A MEMBER OF THE KLUWER ACADEMIC PUBLISHERS GROUP

DORDRECHT / BOSTON / LANCASTER

Library of Congress Cataloging in Publication Data

Carrick, Paul, 1946–
    Medical ethics in antiquity.

    (Philosophy and medicine ; v. 18)
    Bibliography: p.
    Includes index.
    1. Medical ethics.   2.   Medicine, Ancient.   3.   Abortion–Moral
and ethical aspects.   4.   Euthanasia–Moral and ethical aspects.   I.   Title.
II.   Series.
R724.C354   1985   174'.24   85-1878
ISBN 90-277-1825-3
ISBN 90-277-1915-2 (pbk.)

Published by D. Reidel Publishing Company,
P.O. Box 17, 3300 AA Dordrecht, Holland.

Sold and distributed in the U.S.A. and Canada
by Kluwer Academic Publishers,
190 Old Derby Street, Hingham, MA 02043, U.S.A.

In all other countries, sold and distributed
by Kluwer Academic Publishers Group,
P.O. Box 322, 3300 AH Dordrecht, Holland.

*Also published in 1985 in hardbound edition by Reidel
in the series Philosophy and Medicine, Volume 18*

Printed in The Netherlands.

For
Carol, Ben, and Jon

Sleep and Death lifting the body of Sarpedon:
Attic red-figured krater, c. 515 B.C.

Medicine heals diseases of
the body; wisdom frees the
soul from passions.
    — Democritus, *Fragments*

Life is short, the art long,
opportunity fleeting, experiment
treacherous, judgment difficult.
    — Hippocrates, *Aphorisms*

For a physician who is
a lover of wisdom
is the equal of a god.
    — Hippocrates, *Decorum*

TABLE OF CONTENTS

PART THREE / ABORTION AND EUTHANASIA

FOREWORD

The idea of reviewing the ethical concerns of ancient medicine with an eye as to how they might instruct us about the extremely lively disputes of our own contemporary medicine is such a natural one that it surprises us to realize how very slow we have been to pursue it in a sustained way. Ideologues have often seized on the very name of Hippocrates to close off debate about such matters as abortion and euthanasia — as if by appeal to a well-known and sacred authority that no informed person would care or dare to oppose. And yet, beneath the polite fakery of such reference, we have deprived ourselves of a familiarity with the genuinely 'unsimple' variety of Greek and Roman reflections on the great questions of medical ethics. The fascination of recovering those views surely depends on one stunning truism at least: humans sicken and die; they must be cared for by those who are socially endorsed to specialize in the task; and the changes in the rounds of human life are so much the same from ancient times to our own that the disputes and agreements of the past are remarkably similar to those of our own.

The process of medical history and medical technology has hardly made the concerns of ancient medicine irrelevant or even largely outmoded; and even the changes in the perception of the physician's power and responsibility from the Greek world to our own confirms the pertinence of that world's thoughtful review of its own medical arts. Perhaps the puzzling Hippocratic injunction against surgery is particularly suggestive in this regard. It should remind us how disputatious the themes of Greek medical ethics were, how often authoritative views clashed or were rejected in actual practice, how subtly different were the ancients' notions of health and well-being from our own, and yet how clearly continuous is the merely human sense of the serious responsibility of medicine linking the Europe of the pre-Christian era and our own eclectic age.

Paul Carrick is to be congratulated for the persistence and plain good sense of returning to classical views of medicine focused for relevance by a grasp of the comparatively recent new flowering of public and professional disputes about the principal ethical questions. Our own age is distinguished by a remarkable increase in average longevity, a remarkable decline in infant mortality, a remarkable increase in the world's population, and a variety of remarkable technological innovations that call into question more and more fundamentally the sedimented assurances of an allegedly perennial medicine — for instance, as in dividing the issue of removing a fetus from the mother's womb and of ending its life, or as in artifactually producing new stocks by genetic engineering, or as in recycling organs by transplant, or as in compartmentalizing the processes of dying between the brain and the heart and lungs, or as in producing environments within which radically disadvantaged and defective offspring can be sustained indefinitely.

Carrick's book is one of the first to undertake the recovery of ancient views specifically within the setting of bringing them to bear on the innovations of contemporary medicine. The labor of assembling the ancient views in a readable, reliable, and referable form is already quite a heavy one: the temptation to swerve into the marvellous intricacies of the theory of humors, say, could easily have subverted the usefulness of his effort to hold everything together compendiously — by simply offering the easy diversion of what could not help being fascinating in its own right. But he has resisted this; and in doing so he has managed to survey, and to offer deliberately as a survey, an extremely orderly and comprehensive discussion of the principal developments in the theory of medicine and morals of the ancient world. We are in his debt for that. Carrick's reliance on Edelstein, Sigerist, Kudlien, Amundsen, and others is scrupulously clear. It could not have been otherwise. What he has accomplished is to construct a convenient foyer of concepts through which we may pass between our own puzzles and those of the ancient world.

This may in fact be the unperceived special virtue of Carrick's effort. He does not attempt to solve the great ethical issues of contemporary medicine by reference to the materials of the Hippocratic Oath or the views of other ancient traditions; and he does not "solve" the problems of the ancients either. Instead, he offers us a very good sense of the long continuum of speculation from ancient times to our own times, focused on those issues that were both dear to the ancients and are close to our own concerns; and he preserves throughout a genuine feeling for the complexity of ancient disputes, the natural complexity of the issues at stake, the changing focus of our own technologized medicine, the native complexity of our own quarrels, and the need to keep viewing them (both through change and as relatively unchanging) within the lengthening tradition of the history of medicine and human values.

It is the custodial importance of the tradition of the theory of medical ethics that is crucial. The pretense that the essential issues are really timelessly settled once and for all and that those who are uncertain about the received wisdom or who deviate need only be brought back into line one way or other is, ultimately, the tactfully identified target of Carrick's careful study. We learn, by learning, of the profoundly historical nature of medical ethics and of the sense in which our orientation must be perennially refashioned and refitted to the features of every changing time — which is to say, neither simply discovered once and for all nor simply dictated once and for all by an explicit convention or agreement. It evolves in a way that catches up our reflexive attention to how the lengthening tradition can be adjusted to continuing change and the way such an adjustment itself shapes a new sensibility and a new tradition. The mystery of this equilibrating process is not really unique to medical ethics, but it surely cannot be ignored there. Our sense of the process is aptly linked to our own grasp, our peculiarly contemporary grasp, of the open-ended possibilities for extending or interpreting in alternative ways whatever one may regard as a genuinely shared tradition or practice.

We therefore see ourselves in the ancients. Their disputes were obviously eminently humane. But they were not resolved by the discovery of Medical Truth. So Carrick's account is a prophecy of our own continuing need to sustain dispute. That we shall never come to the end seems to be the message; but also, that we shall never need to. Still, we cannot claim in the process to have no understanding at all of what the humane options are.

The most striking feature of Carrick's investigation is the intimate connection it preserves between the ancient attempt to address the questions of medical ethics and the spontaneously recognized need, in doing just that, to see solutions defended within larger theories of a philosophically serious sort. This perhaps is what threatens to separate us from the ancients and what the vigorous, recent revival of a practice of high dispute about medical ethics may well redeem. We cannot separate the injunctions of the ancients from their own theories; and we do so for our own questions at our peril.

Philadelphia, Pennsylvania JOSEPH MARGOLIS
February 1984

PREFACE

During the course of researching and writing this book, I have been struck repeatedly by three thoughts in particular. First, given the gravity of the topics it involves, medical ethics is truly everybody's business. As I will suggest, it has been recognized to be so for a very long time. Second. "bio-ethics" — despite its fancy new name — is on the whole old philosophical ethics. That is, it is for the most part constituted by time-honored and familiar ethical principles ranging from Aristotle to Kant and, in our own day, Rawls. These principles are now being usefully pressed into contemporary domains of medical, social, and scientific service with renewed dedication and promising results. And third, the pace and scope of the current biomedical revolution is not likely to redefine radically the basic relationship of mutual trust and personal autonomy between physicians and their patients any time soon. In fact, the closeness of this relationship may even be enhanced in the decades ahead. The dawning realization is that the latest high technology options in medical care, such as the birth of so-called test tube babies like Louise Brown in 1978, and the still experimental, artificial heart received by Barney Clark in 1982, signal a new age of intimacy and hope between doctor and patient quite unrivaled in the history of medicine.

Given these and numerous other dramatic breakthroughs in health-care during the twentieth century, it is fair to ask: Why bother to explore the origins of Western medical ethics? To some, at least, it appears as if a host of lastingly important developments in both medicine and morals are uniquely occurring here and now. If so, who can afford the luxury of mounting the carrousel of history for yet another ride full circle from past to present?

Having directed this volley of needling questions at myself over four years ago when the present inquiry began, my answer has remained the same: Who can afford not to?

Consider, if you will, these observations — all of which are symptomatic of a lost sense of history:

1. We live in an era when much of the educated public still professes an almost reverential admiration for the ethical teachings of Hippocrates. At the same time, much of this public privately, and often unaware of their inconsistency, strongly disagrees with the Hippocratic Oath's prohibition of abortion and euthanasia as acceptable medical alternatives.

xvii

2. We live in an era when fresh attempts to define what medical ethics is all about frequently ignore the moral principles of this subject's humble beginnings. Yet new ethical theories are sometimes little more than paraphrases of traditional Hippocratic insights dressed up in the latest contemporary jargon.

3. We live in an era when a good many progressive journalists, clergy, lawyers, philosophers, physicians, and social scientists of every background are displaying an uncommon eagerness to christen themselves specialists in bioethics. Yet not a few are doing this, sad to say, as if the entire subject came into being for the first time around 1958!

In view of the preceding evidence, there is every reason to suppose that any time devoted to recovering in a critical way the ancient foundations of medical ethics will be time well spent. Indeed, to lose our clear sense of this rich story of our bioethical heritage may cause us to stereotype the past and, worse still, under- or overvalue our humanistic and scientific accomplishments in the present. In my opinion, this is a brand of self-deception we can ill afford.

I have been greatly inspired by, and am much indebted to, the studies of ancient Western medical values, practices, and traditions that were undertaken in the earlier decades of this century by Ludwig Edelstein, W. H. S. Jones, Henry Sigerist, and, most recently, Darrel Amundsen, William Frankena, Fridolf Kudlien, and Owsei Temkin. In addition, I have benefited immensely from the constructive criticisms and lively discussions of my former teacher and friend, Ronald Hathaway. My father, Louis Carrick, a biologist from the old school with a solid footing in the humanities, offered much encouragement throughout; as did my uncle, Lee Carrick, a physician I have long admired. Philip Lockhart rendered invaluable assistance with some of my initial translations and helped me negotiate the unseen perils of Scylla and Charybdis faced by all who navigate the coastal waters of classical antiquity trained in philosophy only. Elizabeth Beardsley, K. Danner Clouser, Michael Dockery, Joseph Margolis, Norman Vanderwall, Philip Wiener, and William Wisdom read all or portions of earlier drafts and contributed several sound suggestions. So, too, did the Philosophy and Medicine Series co-editors, H. Tristram Engelhardt and Stuart Spicker, plus several unnamed reviewers for Reidel. One of these reviewers, I now know, was Darrel Amundsen. To him I owe a special debt of gratitude for his many detailed and perceptive comments. I cannot thank these individuals enough for the care and attention they have given to this project. Any flaws that may remain, despite their good counsel, must be blamed entirely on me. The first-class typing services of Barbara McDonald were literally indispensable.

To my wife, Carol, and our children, Benjamin and Jonathan, I warmly dedicate this work.

Camp Hill, Pennsylvania   PAUL CARRICK
January, 1984

ACKNOWLEDGMENTS

Grateful acknowledgment is made to the following authors and publishers for permission to reprint selections from copyrighted material:

From *Ancient Medicine* by Ludwig Edelstein. Copyright © 1967 by the Johns Hopkins Press. Excerpts reprinted by permission of the Johns Hopkins Press.

From *Ad Lucilium Epistulae Morales* by Seneca. Translated by Richard M. Gummere. Copyright © 1962 by Harvard University Press. Excerpts reprinted by permission of Harvard University Press.

From 'Principles of Medical Ethics' by the American Medical Association. Copyright © 1980 by the American Medical Association. Reprinted by permission.

From 'A Patient's Bill of Rights' by the American Hospitial Association. Copyright © 1973 by the American Hospital Association. Reprinted by permission.

From the 'Declaration of Geneva' by the World Medical Association. Copyright © 1983 by the World Medical Association. Reprinted by permission.

From the Metropolitan Museum of Art, New York City, for permission to reproduce the Euphronios krater (1972.11.10) used in the frontispiece.

INTRODUCTION

This book explores the origins and development of medical ethics as practiced by the physicians of ancient Greece and Rome. In it I indicate the relevance of their ideas to medical thought today. For example, I shall be identifying and interpreting what many of the leading Greek philosophers and physicians thought about abortion and euthanasia. The related topics of infanticide and suicide are also examined. Every effort is made to furnish a clear picture of how these life and death questions were regarded by physicians, by their patients, and by the philosophers who sometimes made their moral judgments of these matters public. It is hoped that through this investigation the reader — whether a generalist, specialist, or layman interested in the problems of medicine and morals — will gain a more comprehensive understanding of just how deep the antecedents of contemporary medical ethical problems extend into the distant centuries of our scientific and humanistic heritage.

Among the questions to be explored are these. Why did some philosophers and not others judge the life of the fetus to be worth preserving? What value did the philosophers and physicians of antiquity accord to the life of the chronically or terminally ill patient? When was human life judged to begin and end? Was death, according to their religious persuasions or philosophical speculations, something one should rightfully fear?

Today, as in earlier times, it is widely known that the Hippocratic Oath disowned acts of abortion or euthanasia. But what has become increasingly clear just within the twentieth century is that, on the whole, Greek and Roman doctors ignored the Oath's opposition to these measures. This raises the question of whether the pro-abortion and pro-euthanasia positions taken by several leading ancient philosophers contributed in any significant way toward influencing the moral thinking of physicians and others who resisted the Oath on these matters.

My conjecture is that philosophical opinions did play a role, then as now, in influencing both public and professional attitudes concerning the rightness or wrongness of abortion, infanticide, euthanasia, and suicide. I gauge this philosophical influence in Part Three of this study. Pythagoras, Plato, Aristotle, and the Roman Stoic Seneca constitute the center of my discussion there. But, in addition, relevant Greek and Roman legal developments as well as the powerful influences of popular morality and religious cult beliefs are also consulted. Thus I aim at reconstructing how the Greeks, and to a lesser extent the Romans, conceived of the doctor-patient relationship in its entirety. This is the broad question to which Part One of this three-part study is specifically devoted. Yet, in a very real sense, it is the looming and unifying question with which each succeeding part is ultimately concerned.

Accordingly, Part One explores the social setting of Greek medicine. It is designed to lay the groundwork for everything that follows. This it does by examining the social and scientific milieu within which ethical judgments on abortion and euthanasia were made. Among the principal questions raised are the following. What socio-economic, religious, or moral constraints did the average fifth or fourth century B.C. Greek physician experience or recognize in his daily practice of medicine? How did these factors shape his attitudes and conduct toward his patients and toward his community? What did the ancient philosophers think of the physicians of their day, and vice versa?

Among my conclusions, I argue that physicians and philosophers often competed for the public's ear, just as they do to some extent in our own modern era: both dispensed advice on the important moral question of how citizens *ought* to live in order to attain well-being. I further suggest that Democritus, Plato, and Aristotle, among other Greek philosophers, talked about preventive medicine issues such as the responsibility of patients for maintaining their own health.

On the matter of death and dying, I show that at least four distinct perspectives on the afterworld are discernible in the philosophical, medical, religious, and imaginative literature of the Greeks. Furthermore, I explore medical, philosophical, and popular perspectives on (a) the moment death is said to occur and (b) attitudes toward the dying process. It becomes increasingly evident that the Greeks held a myriad of views on these metaphysical and eschatological questions, many of which persist in our own pluralistic society. Therefore I caution that care must be taken not to cast the thinking of the Greeks on these subjects into a single, oversimplified mold.

Part Two specifically deals with the birth of Greek medical ethics. Attention is also given to the medical values of neighboring Mediterranean cultures, including Assyria, Babylonia, Persia, and Egypt. The purpose of this comparative treatment is to gather evidence on which to base a more fully informed judgment of the alleged originality of the Hippocratic Oath. In addition, I discuss the mystery of who the historical Hippocrates really was; it remains a curious fact that little is known about the legendary father of medicine.

All segments of the Hippocratic Oath, then, are critically examined within the broadest social context. This includes the enigmatic prohibition against surgery and the covenant binding medical students to their teachers. At bottom, I attempt to settle the question of the Oath's true origins and purpose. The delicate distinction between medical ethics and mere medical etiquette is also traced to the complexities and mixed motives of the ancient medical marketplace, where doctors plied their healing trade at considerable risk to their own reputation and their continued economic well-being.

Part Three examines a variety of divergent Greco-Roman ethical views on abortion and euthanasia. Here, too, literary materials drawn from selected writings of Greek and Roman playwrights and poets, as well as anthropological, linguistic, and legal evidence are introduced into the discussion. My intention is to supplement the philosophical arguments and medical ideas which constitute the focus of Part Three in such a way that significant social developments are never far to seek. These include ancient contraceptive practices and techniques, types and methods of suicide, and the general competition for patients among physicians, midwives, charlatans, and others. In this way I try to avoid the pitfalls of an otherwise vacuous account. In Chapter Eight, I discuss the ancient physician's sense of moral responsibility in dealing with requests of patients for abortion and euthanasia. My discussion centers on the physician's social role as the primary healer of the sick and suffering. It turns out that the dominant medical ethical values expressed in the Hippocratic writings considered as a whole do not unequivocally oppose or condemn physicians who assist in abortion or acts of voluntary euthanasia, as some mistakenly suppose.

Having highlighted some of the principal topics and conclusions of my investigation, I would like to comment on its scope and method.

The scope and my central focus extend from the sixth through the first centuries of the Pre-Christian era. I have chosen this historical period for two reasons. First, it covers the crucial epoch during which both rational philosophy and scientific medicine first took hold on the continent of Europe. Second, though what would now be called medical ethical issues were not among the leading and recurrent topics of debate among Greek philosophers and physicians, medical ethical issues were nonetheless addressed by a number of these ancient authors. Thanks to posterity, we have some evidence of their moral positions on abortion and euthanasia, as well as on infanticide and suicide. In a few cases, what we do not possess by way of direct evidence of their writings on these particular topics, one may venture to infer from some of their comments on related issues. To this I must add that while my initial focus is primarily on the Pre-Christian era, it is by no means rigidly confined to this period. I do not hesitate to extend the inquiry occasionally into the early Christian and Roman centuries whenever a sharper sense of the issues and arguments is thereby gained.

My method of inquiry is both historical and analytical. It is historical in the sense that my account is bound by the extant evidence of Greek literature, both philosophical and medical, that is presently at our disposal. As in almost any historical inquiry, my guiding purpose is to make sense out of the available evidence not only from (a) the point of view of the relevant subjects of that history (in this case, ancient philosophers, physicians, and patients), but also from (b) the point of view of contemporary students of Greco-Roman history, medicine, philosophy, and culture. In approaching these historical materials, I do not assume that one can readily interpret ancient medical ethical problems and practices in exactly the same way that these phenomena were commonly understood by people living during the distant Greek and Roman eras. But I do assume that, given sufficient evidence, an approximate picture of the values and conduct associated with these traditional bioethical problems can be responsibly secured in a manner that enriches our contemporary perspectives on these issues.

It may well be true that the categories and concepts of one's own epoch tend to influence almost any appraisal of the historical evidence — whether the evidence is philosophical, sociological, or anthropological. But this very tendency also invites author and reader alike to consider further the historical evidence in terms that may suggest interesting implications and parallels in regard to our present historical setting, thereby enriching, not subverting, the interpretative process. Such an opportunity, I believe, constitutes one of the most attractive justifications for pursuing almost any inquiry into the history of ideas.

Finally, my method is analytical in the sense that I am seeking to interpret and evaluate selected aspects of the topics before us in a manner that exposes the logical relations among the key concepts that constitute the framework of ancient medical ethics. Specifically, Greek concepts dealing with health and disease, death and dying, forms of human life, and the moral responsibility of the ancient physician are of decisive importance to the aims of this investigation.

PART ONE

THE SOCIAL AND SCIENTIFIC SETTING

CHAPTER ONE

THE STATUS OF THE PHYSICIAN

In order to fully penetrate the values and conduct of the ancient Greek physician, it is necessary to examine first certain fundamental elements of his historical setting. Thus I have selected three areas of initial inquiry which hold the key to comprehending the cultural framework within which the ancient physician practiced his craft. These three areas are best introduced in the form of three questions. First, who was the Greek physician? Second, what were some of the leading theories of health and disease that guided his care of patients? And third, what were the dominant attitudes and theories of his culture regarding death and the afterworld?

In Part One I shall develop answers to these three important questions, with the present chapter restricted to the first among them. My primary focus throughout will be centered on cultural developments in and around the Greek mainland between the sixth through fourth centuries B.C. My secondary focus will include selected features of the Greco-Roman world up to the time of Galen in the second century of the Christian era.

TYPES OF PHYSICIANS

Who was the ancient physician of fifth century Greece? What was his social rank? Ludwig Edelstein has pointed out that "whatever the situation may have been in prehistoric Greece, in Homer's time [c. 850 B.C.] the physician was already the member of a lay craft; he was 'a worker for the common weal' like the maker of spears, the singer, the seer. He remained throughout antiquity a craftsman, an artist, a scientist."[1] As such, like all Hellenic craftsmen (*demiourgoi*), the physician was seen to be in the possession of the technical body of knowledge peculiar to his craft alone. He was not ordinarily confused with other tradesmen. Nor was the physician equated in the public's mind with the exorcist or magician, who were common enough figures in the ancient world. Hence, by the fifth century one is hardly surprised to find the playwright Sophocles asserting that "It is not a learned physician who sings incantations over pains which should be cured by cutting."[2] Centuries later, Roman law was to establish that any physician who practiced by exorcisms or incantations could under no circumstances sue

3

his patients for unpaid fees. For such methods were judged not to belong to the *bona fide* craft of medicine.[3]

Nor is it likely that medicine in Greece originated in the temples of the deified hero-physician, Asclepius. By the time Asclepius rose to the status of a god and temples were erected to celebrate his powers and promote his cures (most notably at Epidaurus), the physicians (*iatroi*) of Greece were already firmly established as independent, secular craftsmen.[4] The honorary title of "Asclepiad" by which physicians were sometimes addressed had been a source of longstanding confusion among historians attempting to sort out the cultural roots of the Greek practitioner. But it is now emerging from the available evidence that this title did not refer to the god or to a formal religious corporation.[5] Rather, it referred to certain guilds or families of physicians who handed down medical knowledge from father to son, or master to apprentice.[6] The title of Asclepiad was, it seems, a well-chosen indirect reference to the admirable deeds and personal compassion associated with the heroic founder of medicine, Asclepius. The legacy of Asclepius was such that contemporary fifth century physicians sought a share in it as distant descendants or "sons", as they sometimes called themselves.[7] Indeed, few physicians wished the public to lose sight of Homer's oft repeated judgment that "a healer is a man worth many men . . . ."[8]

But while the physician of Hippocratic times (roughly 450–300 B.C.) was, by definition, one who engaged in the art of preserving or restoring health, evidence exists in the writings of the philosophers Aristotle and Plato which invites the judgment that the social status of physicians varied significantly depending on their training and methods. Aristotle, for example, says that the term " 'physician' means both the ordinary practitioner, and the master of the craft, and thirdly, the man who studied medicine as part of his general education."[9]

It is hard to know whether Aristotle here intends a faithful description of well-recognized social groups within Greek medicine as such, or whether he is supplying us with a personal evaluation of the sorts of physicians one usually encountered in the Greek city-state. I think he is rendering an evaluation. The context of his remark is a large discussion in which he is seeking to evaluate how much power a political majority should exercise over the technical affairs of the state. His mention of levels of skill among physicians is thus seen to be part of a useful analogy: just as some physicians do not practice their art but acquire it as part of their general education and thereby know something about medicine, so too many citizens know something of the affairs of the state though they do not personally practice the art of rulership. Also, when one considers the medical knowledge possessed by such Pre-Socratics as Empedocles as well as by the later sophists and generalists of Greece and Rome, it is not at all difficult to account for Aristotle's third sort of generally educated physician.

But then what of Aristotle's remaining distinction between "the ordinary practitioner" as opposed to "the master of the craft"? Does this refer to two distinct social classes of physicians, one the scientist-theoretician, the other the leech-practitioner?

The plausible answer is that it does not. At most it attests to Aristotle's awareness that some physicians, like Diocles of Carystus, appeared as masters of their craft in the sense that they could *explain why* their treatments worked. Others could not, though they still may have gotten helpful medical results. The former were masters, the latter ordinary practitioners. This difference in the practitioners' level of understanding, at once both detectable and significant for a theorist of knowledge like Aristotle, almost certainly constituted the basis for his threefold grouping here. This interpretation is all the more credible since in fact anyone in the ancient world could practice medicine; there were really no formal educational class distinctions as such. To practice medicine was a right, not a privilege. Physicians were not in the least certified or categorized by the state.

Furthermore, it is pertinent to mention that Plato distinguishes two social ranks of physicians in his *Laws*. He speaks on the one hand of the free-born physician, and on the other of the slave-physician. The free-born physician "for the most part attends free men, treats their diseases by going into things thoroughly from the beginning in a scientific way"; the slave-physician generally treats fellow slaves and never gives a patient any account of his complaint.[10] He is "innocent of the theory", whereas his free counterpart can be found "talking almost like a philosopher, tracing the disorder to its source, reviewing the whole system of human physiology . . . "[11]

Yet despite Plato's words, it would be overhasty to conclude that the social difference between slave and free doctors made a decisive difference in the quality of their medicine. Owsei Temkin points out that slave doctors were a known institution throughout antiquity. Highly educated persons could become slaves when they or their defeated armies became captives. So there is little warrant for supposing that physicians belonging to this slave class were automatically of inferior skill or knowledge.[12] In view of these facts, I am prepared to conclude from the *Laws* that Plato, like Aristotle, felt it instructive to distinguish those healers who studied nature in such a way that they deduced their treatments from universal principles, from those less rigorous healers who acquired their skills mostly by imitation and rote. The latter were bereft of a theoretical causal framework for their medicine. The former alone were "almost like a philosopher" and stood for both men as exemplars of their craft.

Perhaps the safest conclusion to be drawn about the status of the Hellenic physician, one fully consistent with the above testimony, is that his personal status was variable. Naturally, if his cures were known to be successful and if he thereby gained a reputation as a highly capable practitioner, he would enjoy a measure of fame and status exceeding that of his more average competitors. Then as now there existed the country doctor, the city doctor, and the regal court physician. When the Greek physician Galen was practicing in Rome in the second Christian century, he complained that physicians treated their patients differently depending on the physician's origins, his training, and the perceived social standing of a given patient. Patients, in turn, reacted differently toward their physicians depending on the physician's perceived authority and his high or low reputation.[13]

But as to the *average* physician taken as a stereotype of fifth century Greece, particularly Attica, I am in full accord with Edelstein's characterization:

> The Hippocratic physician is a craftsman. As a craftsman, he practices either as a resident or as an itinerant; he may also settle for a while in some town, leave again, work in another town, or wander all over the country. When he is in a town, he works in his shop or in his patient's home. The shop is a place in which today one person, tomorrow another, plies his trade – not a hospital, not a consulting room in the physician's house. Sick people come to the shop for examination and treatment, or the physician goes to his patient's home. An itinerant physician works in the patient's house or has a booth set up in the marketplace of the town, or elsewhere, and practices there.[14]

Elsewhere Edelstein has aptly summarized the social situation for doctors down to the time of the Roman Empire:

> About the basic social situation there can be no doubt. The medical practitioner, working for a livelihood and working with his hands, preparing drugs or performing operations, was to the ancients a craftsman and, as such, belonged to the lower strata of the social order. And this was true throughout antiquity. It did not help his standing that in the Classical and in the Hellenistic era he was usually a migratory worker, an 'out of towner,' and in Roman times a foreigner; in the Rome of Cicero no citizen had yet gone into medicine, while he would not scruple to become a lawyer. Nor should one forget that in the empire probably more physicians were slaves than had even been the case in Greece.[15]

Finally, it is relevant to ask: was there anything like our modern medical specialties in classical antiquity? The simple answer is no. For there was no profession as such in today's terms, and no board certification or other licensing system. But practically speaking, it may be admitted that while there was no division between medicine and surgery (the latter then including operative surgery, setting bones, reduction of dislocations, and cauterization), evidence suggests that *some* physicians may have practiced surgery exclusively.[16] But if so, the trend toward specialization was for the most part a matter of personal convenience rather than based on any deep-seated ideological split within the craft of medicine, as some have assumed.[17]

EGYPTIAN INFLUENCES

In contrast to the comparatively unified approach to healing characteristic of fifth century Greek medicine, the Greek historian Herodotus writes of the Egyptian physicians that "medicine is practiced among them on a plan of separation; each physician treats a single disorder, and no more: thus the country [Egypt] swarms with medical practitioners, some undertaking to cure diseases of the eye, others of the head, others again of the teeth, others of the intestines, and some those which are not local."[18] This account raises the question: to what extent was Greek medicine influenced by the Egyptians?

To begin with, Herodotus' report of Egyptian specialization suggests that the state of the medical art in Egypt was quite advanced by ancient standards. In antiquity Egyptian physicians were much sought after in foreign lands. Indeed, the Greeks expressed admiration toward Egyptian medicine, particularly their effective use of drugs. A rich variety of these is prescribed throughout the best preserved Egyptian medical papyri now in our possession.[19] Moreover, we know for a fact that the Greeks copied down many of these Egyptian pharmaceutical formulae. They also imitated and further developed the Egyptian techniques of repeated and careful empirical examination of the patient, including both diagnosis and prognosis, which were developed by the Egyptians perhaps as early as 3,000 B.C.[20] Therefore, both the testimony of the Greeks themselves and the Egyptian medical texts provide a basis for the judgment that Greek medicine was influenced by Egypt.

But how much? The jury is still out on this question. No exhaustive study has been undertaken. Nonetheless, opinions have been offered. The great medical historian, T. Clifford Allbutt, writing in the first quarter of this century, locates the Egyptian influence mainly in drug therapy. He discounts Egyptian anatomy as "rudimentary and pettifogging" even by Hellenistic standards. But he admits that the doctrine of the four humors may have originated in Egypt.[21]

A more generous estimate of the Egyptian contribution is provided by J. B. Saunders. Saunders credits the Egyptians with no fewer than five major accomplishments, the importance of which was not lost on the Greeks. These include: (1) the recording of reusable empirical data on the application of drugs; (2) an unusually effective causal theory of disease based on the putrefaction of digested substances within the body; (3) the development of various purgatives as an effective treatment against internal infection and swelling; (4) pregnancy and sterility tests for women; and (5) the development of a birth prognosis according to which the sex and condition of the developing fetus were forecast.[22] Thus Saunders appears willing to credit the Egyptians with a causal theory of health and disease which recognized a rational framework largely divorced from magic and supernatural explanations.

To me this seems overgenerous, based on the evidence of the five principal Egyptian medical papyri currently under study. Personally, I think that the most one is entitled to say is this. While there are *some* empirical-rational elements to be found in the papyri (especially reasoning by analogy and going from specific cases and symptoms to general treatment rules), alongside these pre-scientific techniques one cannot ignore the Egyptians' reliance on magical incantations and supernatural models of disease.[23] Disease is attributed to the workings of an invading, incapacitating deity or spirit of some sort. Therefore, I am inclined to credit the Egyptians with their advanced pharmacopeia, the law-like rational elements in their theory of putrefaction, and their sharp clinical observations based on the methodological strategy of diagnosis and prognosis, as having plausibly influenced Greek medicine. This much seems justified based on the evidence.

But to the Greek physicians of the fifth century alone must go the credit for divorcing medicine from magic — something the Egyptian physicians failed to do even in Herodotus' time.[24] Thus one must not be over-enthusiastic in ascribing the beginnings of scientific medicine to Egypt without at least this very important acknowledgment.

THE PROBLEM OF LICENSURE

Yet one feature common to both Greek and Egyptian cultures was the absence of a system of medical licensure.[25] While little is known about the regulation of the medical profession under the Egyptian pharaohs, it is clear that compared to the freedom enjoyed by their Greek counterparts on the islands of Cnidus, Cos, and at Crotonia (in Sicily), the Egyptian physician was required to practice according to very rigid rules of treatment. One reason for this was the widely held opinion in Egypt that it was hard to improve upon the medical wisdom of older times. Hence the first century B.C. author Diodorus Siculus, who based his opinion on earlier Egyptian documents, wrote that Egyptian physicians gave treatment in accordance with strict written procedures. If their patients died, such physicians were usually absolved from any charge. But if they deviated from the traditional practices and methods in any way, they were subjected to the death penalty.[26] Aristotle in his *Politics* describes a slightly more flexible situation: "In Egypt the physician is allowed to alter his treatment after the fourth day, but if sooner, he takes the risk."[27] The risk to which Aristotle refers probably includes the death penalty, especially in the event the patient dies.

In comparison, the Greek physician was virtually immune from civil penalties if his patient died while under his care. Of course, this did not mean that he could do anything he liked to his patients. For example, he could not wantonly murder his patient and be expected to escape prosecution for homicide after charges were brought by relatives of the deceased. But he practiced his craft by and large without interference and was free to treat as best he knew how. The freedom that he enjoyed — plus the opportunity for possible harm to others as a result of this enviable freedom — did not escape the attention of his own countrymen, either. Thus, according to the comic poet Philemon, the Greek doctor is the only person permitted to kill and not die for his crime. He is the only citizen to enjoy complete immunity for causing the death of others, the Roman Pliny was later to write. For in Roman times the physicians were just as free and perhaps even more taunted by their critics than they had been in Greece.[28]

Just how this freedom from licensure and state control led to the gradual development of medical etiquette and ethical guidelines will be addressed in Part Two. But the point to be made here is that the freedom enjoyed by the Hippocratic physician was at once both a personal burden and an occasional source of scientific experimentation and discovery. That is, on the one hand he may have been tempted on occasion not to take a difficult case for fear that, if the unfortunate patient died, this might get around town and hurt his reputation — which was effectively his only employment credential. This sort of gloomy gossip could have been bad for business. On the other hand, unlike the Egyptian practitioner, the Hippocratic physician was free to make adjustments in his treatment of a disease as he saw fit. This he could do at any time without civil threat or penalty. It meant that he felt free to experiment to a limited extent on the particular cases before him. He could thus modify established general treatments to better suit the special characteristics of each individual. This freedom to improvise and improve treatment constitutes one of the fundamental strengths of Greek medicine. It helps to explain its essentially progressive and diverse nature as well as its eventual need for *voluntarily* followed codes of conduct. These codes originated from the medical practitioners themselves, and were not (as was the case in Egypt) imposed by the state.

Thus, several of the features of the social setting in which the average Hellenic physician worked have come into view. I have argued that he was regarded as a craftsmen. As such, like all Greek craftsmen he was free to set his own fees and work for whom, when, and how he chose. He was often a stranger, wandering from village to town selling his services. Since he worked with his hands, and since medicine was excluded formally from the ancient list of liberal arts, his social status was comparable to that of other common workmen or craftsmen (*technai*), and generally below that of musicians, mathematicians, poets, and philosophers.[29] In a word, he was a businessman. In the absence of any system of licensure, he needed to protect his reputation as a wise and useful healer of the sick. Otherwise he would be confused with charlatans and quacks, of whom there were many. As I have suggested, compared to his Egyptian counterpart, the Greek doctor was freer to regulate his practice and adjust treatments as he saw fit. While usually pious in his attitudes toward the gods, he was not in the habit of attributing disease to supernatural causes. Nor did he resort to magic, as did Egyptian physicians and priests (often the same person) in Pharaonic times.

Yet on balance it will also be recalled that the Greeks possessed a well-deserved reputation among the ancients for advocating the importance of a sound mind in a sound body (*mens sana in corpore sano*). Health was among their most prized values. As such the physician, *qua* the restorer and preserver of health, was no ordinary craftsman. If he was known to be effective in his cures, he was indeed much sought after. This is why it must be underscored that, in the final analysis, the social status of a particular doctor *varied* with the esteem in which his public regarded him. His performance and resulting reputation were decisive to his financial success and his ability to eke out a living.

If a dct in greece actually cured people he was looked at w/ much social status, otherwise he was like anyone else.

THE RELATIONSHIP BETWEEN PHYSICIANS AND PHILOSOPHERS

Unlike our own time, the physician in antiquity was not regarded by his often skeptical public as the first and last source of medical wisdom and authority. Some will be surprised to learn that by the fifth century B.C., well into Roman times,

... everybody was familiar with medicine – it was the only art or science about which everybody knew something; physicians wrote books for the general public; laymen discussed medical problems with their physicians; in short medical knowledge was perhaps diffused more widely in Greek and Roman times than in any other period of history ...[30]

Philosophers, of course, exploited this common fund of public interest in and knowledge of rudimentary medical techniques for their own ends. When they wished to emphasize the peculiar tasks of the philosophical enterprise or instruct others on the nuances of their own theories, the idiom and examples of medicine often proved to be useful pedagogical tools. Hence the atomist philosopher Democritus stated: "Medicine heals diseases of the body; wisdom frees the soul from passions."[31] Aristotle, himself the son of a physician, knew well the persuasive power of the medical analogy in setting forth his ethical prescriptions. The germ of his ethical doctrine of the mean is found early on in Book Two of his *Nicomachean Ethics*. There his concept of virtue is introduced in this striking medical parallel:

... [I]t is the nature of such things [human actions and states of character] to be destroyed by defect and excess, as we see in the case of strength and health ... ; both excessive and defective exercise destroys the strength, and similarly drink or food which is above or below a certain amount destroys the health, while that which is proportionate both produces and increases and preserves it. So too is it, then, in the case of temperance and courage and the other virtues.[32]

In addition, philosophers associated with the ancient Academy were especially eager to coax the public into believing that they ought to care about the health and state of their invisible soul as diligently and thoughtfully as they already cared about the health of their visible body. Aristotle shared a similar concern for the well-being of the soul. Hence Werner Jaeger seems to me quite correct in arguing that Aristotle used the medical model to full advantage in demonstrating the details of his ethical system, as the preceding quotation illustrates.[33]

As to the related question, Who held more prestige in the public's mind, the physician or the philosopher?, there seems little room for doubt. The philosopher was by far more generally esteemed. Yet it must not be forgotten that the average Athenian held in high regard the practical applications of the medical craft. Health, after all, was regarded by many of the ancient Greeks and Romans as the highest good (*summum bonum*) of life; or at least, in Aristotelian terms, many regarded health as the highest external good (that is, it was necessary to the good life). This high value accorded to health suggests that the more famous and successful physicians could wield considerable power and authority. When these successful physicians spoke, people were inclined to listen. There is little reason to doubt that on some occasions, at least, philosophers and physicians competed for influence and authority before their public on moral matters relating to *the right way to live*. For this is precisely the area that the physician is likely to address in prescribing and justifying his regimen for the preservation of health. Yet the acquisition of sound habits and daily regimen is also the proper domain of the philosopher. The philosopher seeks to define what a life worth living is, and may prescribe rules of conduct for achieving it.

Mindful of this potential competition for influence between physicians and philosophers on matters of personal conduct, philosophers from Plato and Aristotle on sought to reassert the supremacy of their discipline as the only proper authority on the ultimate questions of human value. This tendency to draw the line can be seen in Plato's dialogue, *Laches*. At one point in that dialogue, Nicias (with Socrates' approval) asserts that Laches had just spoken falsely of physicians

> . . . because he [Laches] thinks that the physician's knowledge of illness extends *beyond* the nature of health and disease. But in fact the physician knows no more than this.[34]

Elsewhere in the dialogues (e.g., *Republic*, Book Four) Plato tends to reduce the authority and role of the physician by restricting medical assistance to cases of an emergency nature only, such as the repair of broken limbs, gashes, wounds, etc. Moreover, the physician is forbidden to instruct the citizens of Plato's ideal state to seek or preserve their personal health at any cost, or by any means. Such vain advice would not serve the economic interests of the state. It would instead threaten to extend human suffering beyong the limit of a useful, worthwhile life, in Plato's opinion.[35]

To sum up, the relationship between ancient philosophy and ancient medicine is both resourceful and contentious. On the one hand, the philosophers benefited from the richness and power of medical similes and analogies, casting these as useful instructional devices. On the other hand, the physician, particularly if he was renowned or well respected by the public, posed a

potential threat to philosophers by trying to tell others how to live, i.e.,
which medical regimens were *most valuable*. The point that Plato, Aristotle,
and their successors tried to make is that at most the physician is entitled
to *describe* how one might — should one so choose — regain or preserve one's
health. The physician is not entitled to *prescribe* whether one ought to try to
recover or preserve one's health at any cost. In other words, the positing of
ultimate aims and values was seen by most of these ancient philosophers as
a task belonging to the province of philosophy alone. This task did not
properly belong to the medical craft or with any other discipline for that
matter.[36]

CHAPTER TWO

THEORIES OF HEALTH AND DISEASE

Whatever benefits came to philosophy by way of Greek medicine, Greek medicine also received from the early Pre-Socratic philosophers the impetus to emancipate the healing arts from magic and superstition. In this chapter it is my aim to explore some of the main facets of two theories of health and disease that most strongly influenced Greek physicians, namely, the humoral theory associated with Hippocrates and the eclectic theory of the best constitution associated with Galen. The Pre-Socratic influence toward seeking the universal causes of phenomena in the *natural order* governed by lawlike, non-magical forces (knowable to man through rational reflection and observation) was an influence unquestionably felt by the Hippocratic authors. However, I shall not stop to argue this point owing both to the long digression it would entail and to the fact that scholars like Sigerist, Jaeger, and Edelstein, among others, have admirably secured this position and stand virtually unopposed. I shall be concentrating instead on the results of the general Pre-Socratic quest for natural explanations; particular Pre-Socratic philosophers will receive only the briefest mention, and some familiarity with their theories is taken for granted. Also, to promote continuity in the account I am about to give, Democritus, Plato, and Aristotle will receive additional exposure so that their theories of health and disease may at least be glimpsed in relation to pertinent medical developments.

THE HIPPOCRATIC HUMORAL THEORY

It is important to recognize at the outset that while the doctrine of the four humors is widely presupposed or explicitly mentioned by the authors of the seventy or so medical treatises that comprise the Hippocratic Collection, the basic theory about to be explored acquires various interpretations and applications depending on the particular treatise one consults. That is, there was no single, primitive humoral theory explicitly endorsed by the Hippocratic authors. Instead one finds variations and elaborations of a recurring number of limited diagnostic, prognostic, and therapeutic principles which are often systematically linked.

Moreover, although the related question of who this person called

Hippocrates really was will be pursued in the next chapter, it seems helpful to mention here that some of the evidence indicates that Hippocrates may not have personally embraced the humoral theory. Rather, according to the testimony of Aristotle's pupil, Meno, there is reason to believe that Hippocrates himself laid disease to the generation of putrified air which arises inside the body when the body fails to properly digest food.[1] Even so, the medical authors of the treatises with which his name is associated were influenced by Hippocrates' organistic principle according to which one cannot truly understand the nature of part of the body without first understanding the nature of the whole body. Therefore, although Hippocrates himself cannot be definitely linked with the humoral theory *per se*, it seems fair to conclude that the systematic and holistic character evidenced by that theory would bave been congenial to his medical outlook.

As to the humoral theory itself, I have modified Henry Sigerist's splendid account, based on Hippocratic sources, by casting the fundamentals of the humoral theory into seven basic principles which I believe more fully capture the theory's explanatory power.[2] These seven principles, in the order in which they shall be discussed, are

   (i) the principle of equilibrium
   (ii) the principle of seasonal influence
   (iii) the principle of contraries
   (iv) the principle of innate heat
   (v) the principle of natural healing
   (vi) the principle of *pepsis*
   (vii) the principle of critical days.

To begin with, the *principle of equilibrium* is the most fundamental as it asserts that health is the result of a proper balance or equilibrium within the body, whereas illness is caused by humoral imbalance. But what are humors exactly, and what substances constitute them? A treatise taken by many scholars to be one of the earliest in the Hippocratic Collection, entitled *On Ancient Medicine*, identifies the humors with an *unlimited* number of forces or qualities (*dynameis*) such as heat, cold, dry, moist, sweet, bitter, etc. When the blend of these qualities is perfect, a man is in perfect health.[3] But should poor nutrition or some other cause upset the balance, disease occurs.

Yet there is little doubt that talk of forces or qualities is rather vague-sounding. Accordingly, the author of another treatise, *On Breaths*, drops the humoral modal in favor of an explanation of diseases based on the presence or absence in the body of the correct amount and blend of air inhaled from one's environment and drawn into the interstices of various organs. It would appear that this author sought a somewhat more tangible substance than the vague humoral qualities alluded to above; breath appeared to be real and always essential to sustaining human health.

Gradually, though, the tendency in the Hippocratic Corpus was to reintroduce the terminology of the humors and, *unlike* the author of *Ancient Medicine*, limit them to just four. Thus in *On Diseases*, the humors are identified as phlegm, blood, bile, and water,[4] Moreover, in that same treatise each humor is said to originate in four specific organs: blood from the heart, phlegm from the head, water from the spleen, and bile from the gall bladder. Also in that treatise, it is suggested that disease can be traced to three basic causes. The first cause mentioned is the excess or deficiency of any one of the four humors (usually due to an inadequate or overindulgent diet). Second, diseases stemming from violent causes such as wounds, falls, or, on the other extreme, severe fatigue. And third, atmospheric conditions (season, weather, climate) can dispose people toward disease, about which more will be said later. The same author has it that these three factors act on the humors, making them coagulate or liquify, or undergo some related change. In this way, "the vitiated humor attaches itself to some part of the body and the diseases are usually named after the organ affected."[5]

However, it is in the somewhat later treatise, *The Nature of Man*, that water is replaced as a humor by black bile, which is introduced here for the first time as a normal humoral constituent of the body.[6] Sigerist designates this treatise, attributed by Aristotle to Polybus, as the starting point of the classical theory of the four humors. Along with the remaining three constituents, yellow bile, phlegm, and blood, this theory dominated medical thinking for 2,000 years. In accordance with the first principle, in this treatise perfect health is said to prevail when these four humors are in equilibrium (*isonomia*) with regard to their temperature, strength, and quantity.[7] In addition, when one humor reigns supreme (what the Pre-Socratic Alcmaeon called *monarchia*), isolates itself, or no longer cooperates with the other three, pain may originate in two locations within the patient: (1) the site that the humor leaves may become inflamed; and (2) the site that the humor crowds may swell with pain. Similarly, the sudden or abnormal evacuation of a humor (either inside or outside the body) is said to cause considerable pain.

But what exactly, then, is a humor — at least as it is generally characterized by the Hippocratic authors?

These are not fictitious, not mere principles, but very real. Wound the body anywhere and you will see *blood*. Give a drug that acts on phlegm and the individual will vomit *phlegm*; or [*yellow*] *bile* if you give him a cholagogic remedy. Similarly he will evacuate *black bile* in response to certain drugs.[8]

In sum, humors were characteristically taken to be tangible substances originating in the body largely through digestive and related nutritional processes. Furthermore, they are not merely theoretical terms, for they were thought to have determinate empirical content and be spatiotemporally located within the body. Thus it becomes increasingly evident that the explanatory power accorded to the doctrine of the four humors owed not only to its ability to describe and classify diseases, but also to its capacity to provide physiological, causal explanations.

Let us now turn to the second principle. I shall call it the *principle of seasonal influence*. Quite simply, this principle asserts that the qualities hot, cold, moist, and dry always distributed in the four seasons are linked to the four humors always found in man in such a way that changing seasonal conditions causally determine which diseases each person is disposed to get at any given time. This means that the physician was expected to take into consideration, among other things, the climate and weather conditions when diagnosing and treating his patients. For it was suggested that the seasons in fact had some influence on the four bodily humors.

To illustrate, it was observed that nose bleeding frequently occurred in the spring more than at other times. The reasoning proceeded that blood increased then since it is *hot* and *moist* just like the seasonal conditions of spring tend to be. In contrast, summer, which is characteristically *hot* and *dry*, was thought to give rise to yellow bile in the body. This was suggested since physicians observed more bilious stools then and also observed that the skin often manifested a yellow tinge. Thus the data of observation seemed to confirm the principle of seasonal influence.

Also adding to the wide attraction the humoral theory enjoyed in antiquity was the fact that the four qualities hot, cold, moist, dry, could be linked not only to the four seasons, but also to the four elements fire, earth, water, air, which the Pre-Socratic philosopher Empedocles (495–435 B.C.) had identified as the ultimate constituents of the universe. So that part of the humoral theory expressed by the principle of seasonal influence especially could be easily elaborated to connect man's nature (*physis*) to the nature of the entire universe. Hence this cosmological symmetry of qualitative, humoral, and elemental pairings lent added force to the humoral theory by physically joining the microcosm of man with the macrocosm of nature.

The third principle, which I have dubbed the *principle of contraries*, constitutes a further extension of the theory. It serves to systematize the diagnosis and treatment capabilities afforded by the humoral model by specifying that organs, diseases, *and remedies* also all possess the qualities hot, cold, moist, and dry. In view of this, it prescribes that *contraries should be cured by contraries.*[9] The chart below illustrates the relations between the four humors, the four proximal qualities, the four Empedoclean elements, and the four seasons:

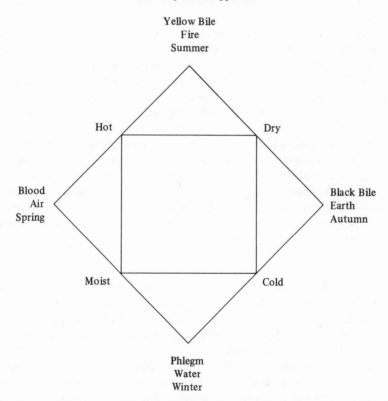

Humoral Square of Opposites

Yellow Bile
Fire
Summer

Hot   Dry

Blood   Black Bile
Air   Earth
Spring   Autumn

Moist   Cold

Phlegm
Water
Winter

What was the logic undergirding this third principle? Why and how the principle of contraries was utilized to aid in diagnosis and treatment may be seen from the following illustration. Suppose a patient complained of abdominal pain. Suppose further that the physician inferred from this complaint

that, on the humoral theory, its cause was probably due to a humoral imbalance associated with the patient's black bile. Black bile was thought to
originate in that bodily organ known as the spleen. Furthermore, when the
black bile was in proper equilibrium with the other three humors, its qualities
of dryness and coldness (see chart, above) stood in a proportion productive
of health. Now, however, suppose the patient evidenced cold chills and suffered further from constipation. These symptoms suggested an excess inside
the body of the qualities cold and dry respectively. The contraries of cold
and dry are, according to the above chart, hot and moist. This information
provided the basis for a treatment plan based on a seemingly rational framework.

Here, in order to remedy the abdominal pain, the patient might be prescribed periodic hot baths followed by a general increase in the intake of
liquids. The aim of this particular remedy, like all similarly devised remedies,
was to bring into proper balance the humoral excesses or deficiencies said to
be causing the affliction. In this case, the hot baths aimed at warming the
overly cold black bile; the increased intake of beverages aimed at moistening
the *overly dry* properties of the black bile.

As useful as the humoral theory was for supplying a causal account of the
many diseases known to ancient man, the four humors alone did not seem to
be able to explain the origin of life itself in the body. Some sort of driving
force was thought to be involved to explain further: (a) How were the humors
developed and sustained by food? (b) What caused them to possess motion?
and (c) What blended them together in varying proportions so as to expel
superfluities and regain balance? This driving force was identified as the
'innate heat' (*emphyton thermon*); it is a vital force that may well have derived from Heraclitus' notion of fire as the embodiment of the dynamic
motion of all existing things. In any case, innate heat was said to have its seat
in the left ventricle of the heart.[10] These considerations introduce our fourth
principle, the *principle of innate heat*, which states that the origin, nutritional
processes, and dynamic motion of the four humors themselves are produced
by the body's innate, fiery heat.

The beauty of this seemingly *ad hoc* principle within the humoral theory
is that it furnished a convenient explanation for several other bodily phenomena that seemed to demand explanation. For example, the rapid heartbeat
and higher bodily temperature of infants could *ex hypothesi* be understood
as the natural result of a higher level of innate heat required to promote
faster bodily growth. In contrast, that older persons have lower bodily
temperatures could be explained since they are through growing and their

innate heat is at a lower level. This fact also explains why older persons generally need less food than the young, i.e., their humors are fully developed and need only to be sustained. Just as neatly explained by the presence of the heart's innate heat is the whole process of respiration. It was held that the fundamental reason one must breathe is that the heart needs to be cooled lest it be damaged by excessively hot temperatures.

Having established that innate heat is an essential part of man's nature, the fifth *principle of natural healing* follows from the yet further consideration best introduced by the Hippocratic author of the treatise *Epidemics*. "Nature," he writes, "is the physician of diseases. Nature finds ways and means all by itself, not as a result of thought."[11] This observation led finally to the fifth principle which states: (a) that nature is a cooperative force which works through each person to bring the diseased body back into that often delicate balance which defines health; and (b) the task of the physician is to understand the workings of nature, to prescribe only those therapies that will support nature, and do nothing that might impede the natural healing processes.

There are two immediate implications of this fifth principle. The first is that nature was not seen to be inherently inimical to man's survival but rather was inclined to be supportive of life. This was deduced from the fact that sometimes very sick or injured patients fully recovered on their own without seeking medical treatment. In this sense also, many Greeks tended to view nature in a positive light and sought to live in accord with nature's laws (especially the Stoics). Second, as nature was seen to work in a regular, sometimes cooperative way, the physician sought to educate himself to nature's regular healing processes. Only then could he be effective as a co-healer and promoter of this wondrous healing power of nature (*vis medicatrix naturae*). The key to healing was a sound medical education; the best teacher was ultimately nature.

This last principle also helps explain why the prognosis of the Hippocratic physician was so important. To know what the likely cause and outcome of a disease would be had much to do with helping promote the natural bodily processes in the right way, *at the right time*. But if the patient was judged beyond help owing to nature's irresistible natural course (which included, finally, decay and death), there was little point in endeavoring to begin treatment other than keeping the patient comfortable. About this decision "not to treat" more will be said in the following chapters.

The sixth and seventh principles of the humoral theory are closely related. Thus it will be profitable to discuss them together. Both are concerned to explain and predict the detailed physiological processes that cause disease in somewhat more precise terms than the five preceding principles do. The sixth *principle of pepsis* states that most diseases are caused by one humor dominating the others in an abnormal, excessive way; the excessive matter of the dominant humor naturally undergoes a thickening process called *pepsis* (or coction), culminating in its expulsion from the body so that humoral balance is restored.

Just what causes one humor to dominate the others in the first place is sometimes hard to specify. But the Hippocratic authors usually ascribed the imbalance to (a) improper diet; (b) lack of proper exercise; (c) bodily injury; or, more generally, (d) an incorrect mode of living in relation to the seasons of the year. At any event, the excessive matter of the offending humor (later called the *materia peccans*) needed to become ripe in order to be expelled. Sigerist elaborates:

The body mobilized all its defensive forces and attacked the raw, faulty humor which under the influence of *innate heat* became 'cooked,' ripe, and ready to be expelled . . . . The expulsion was quick in the form of a *crisis*; or slow and gradual − in the later case we still speak of a *lysis*. The crisis is defined in the treatise *On Diseases* as a condition in which a disease takes a turn for the worse or for the better or changes its character or ends by being healed.[12]

Sigerist has in my view aptly termed this turning point the "decisive battle" between man's nature and the disease. Moreover, one need not look far to realize that this process called *pepsis* or thickening, which led to the expulsion of the offending matter, would seem to be supported by commonplace observations such as the increase of phlegm, sputum, and other putrid bodily fluids that often accompany a variety of diseases (including the common cold). Therefore, one can again appreciate that the theory embodied in this principle seemed well borne out by the data of experience to which the Greek physicians regularly turned.

That the *crisis* of a particular disease tends to be reached in a specifiable number of days − at which time the disease usually takes a crucial turn for better or worse − was the prognostic principle which I shall refer to as the *principle of critical days*.

One sees in this seventh and last principle an effort to mathematize medical treatment along lines that would have been quite congenial to the Pre-Socratic Pythagoras and may have been inspired in part by his teaching that all phenomena are reducible to number. At any rate, the Hippocratic physician observed that many diseases, e.g., the common cold, pneumonia, malaria, etc., appeared to have their own natural course regardless of who suffers them. If so, the effort to closely record and isolate the days associated with characteristic, marked changes in the patient's condition promised to supply predictive power based on the identification of the patient's current stage in the disease process. It also supplied a basis for classifying diseases based in part on detectible similarities or differences in the duration of the relevant symptoms.

In conformity with this last principle, the author of *Epidemics I* is quite emphatic: "One must pay attention to the crises and know that at such times they decide about the life and death or at least the changes for better or worse."[13] Also, in *Prognostics* we read:

Fevers come to a crisis on the same days as to number, both those from which men recover and from which they die. For the mildest class of fevers, and those appearing with the most favorable symptoms prove fatal on the *fourth* day or earlier; and the most malignant and those setting in with the most dangerous symptoms prove fatal on the *fourth* day or earlier.[14]

One can discern in these words the growing tendency (now a common feature of modern medicine) to mathematize the signs and symptoms of diseases and to define more precisely diagnostic, prognostic, and therapeutic procedures. It is also possible that the emphasis on the number four in these and other passages throughout the Hippocratic Corpus is indicative of Pythagorean influences.

What is more certain, however, is that though there was admittedly a strong speculative element in these attempts to link the critical days of illness to determinate disease outcomes, the bulk of the clinical observations on which such predictions rested were in themselves accurate. Indeed, in this seventh medical principle, as in almost all that have been presented, it is evident that, while from the modern medical perspective the Hippocratic physicians did not know the true causes of most diseases, they often knew the effects. These effects they diligently recorded and classified based on recurring symptom sequences from which they tried, as good empiricists, to reach general conclusions. Moreover, one finds in their inductive, systematic, and mathematical approach the embryonic beginnings of a genuinely rational and scientific medicine.

Having discussed the seven basic principles of the humoral theory, we are now in the position to notice further that it construes disease and health in basically physiological, not ontological, terms. Here a brief digression is necessary. Traditionally there have been two alternatively dominant and

opposing concepts of disease throughout history: (1) the *physiological* and
(2) the *ontological*. The former characteristically (a) construes disease as a
consequence of imbalanced functions occurring within the individual; (b)
emphasizes the uniqueness of each person's affliction; and (c) construes the
pathological disease processes as occurring inside the whole person instead
of ascribing the dysfunction or disease to outside, invading disease entities
(*entia morbi*).

In contrast, the latter ontological concept of disease characteristically
(a) does not construe disease as a consequence of imbalanced functions
occurring within the individual; (b) emphasizes the specific reality and distinctiveness of the disease entity (*ens morbi*) in contrast to the patient
harboring it; and (c) construes the disease as a discrete, usually localized
entity which is external to the healthy person but which, in seating or manifesting itself inside the individual, produces the onset of disease.

Now it has been shown that, on the humoral theory, the diseased patient
as a whole is regarded as sick. In the main, his illness is the result of a humoral
imbalance, a dysfunction involving the entire person. For the humors were
thought to run throughout the body in general, even though the faulty humor
might be more concentrated in one region than another. True, some diseases
were construed as localized on the humoral theory, but only roughly so and
not very often. Hence the Hippocratic authors were disinclined to view disease
as a specific entity with an ontological status independent of the whole patient. For example, one such author describes what we now call mumps as
"swellings in the region of the ears" (*eparmata para ta ota*).[15] But characteristically, he attaches no name to this particular disease, for the disease
does not have any separate status, nor does it have a determinate seat in the
anatomy of the body.

THE DEMOCRITEAN ATOMISTIC THEORY

In contrast with the physiological concept of disease that is associated with
the humoral theory found in the Hippocratic Collection, the ontological
concept of disease according to which diseases are usually construed as discrete, localized entities can be hypothetically reconstructed from the writings
of Democritus (460–370 B.C.) and the later atomists. It will be recalled that
Democritus sought to explain all the observable phenomena of both the
macrocosm and the microcosm as the result of the size, shape, and arrangement of invisible, indestructible atoms afloat in the void. What is less widely
known is that he wrote on medical topics, producing at least three books that

are lost to posterity, entitled *Prognosis, On Diet*, and *On Medical Method*. In this connection, Democritus is reported to have said that one can preserve one's health through self-discipline. He wrote:

Men ask in their prayers for health from the gods, but do not know that the power to attain this lies in themselves; and by doing the opposite thing, through lack of control, they themselves become the betrayers of their own health to their diseases.[16]

Like Pythagoras before him, Democritus stresses that moderation in all things brought pleasure and well-being. Moreover, according to at least one ancient commentator, Democritus attributed mental illness to the excessive hotness or coldness of the fine round atoms that constituted a person's soul.[17] After reporting on Democritus' materialistic theory of illness, Theophrastus (c. 370–285 B.C.) states, "it is evident that he made thinking dependent upon the mixture of the body — which is perhaps reasonable enough in one who makes the soul a body."[18]

Reducing all psychological states to the mechanism of the material soul-atoms uniquely arranged in the void (i.e., the spaces or pores of the patient's body in this context), and construing all illness in the final analysis as the result of and identical with the body's disordered and misarranged atoms, it seems quite fitting that Democritus urged the attainment of an even, cheerful disposition. For such an ethical ideal, if realized, would on his theory keep the soul atoms free from the sort of *physical agitation* that he associated with the violent passions. These violent, angry passions, having a material basis, could in turn produce unhappy, deranged thoughts. This could so affect the individual as to render him mad.

But the point at issue here is that, whether a disease is termed mental or physical in our conventional usage, for Democritus all diseases are at bottom physical. Thus on his view all diseases were thought of as having a determinate seat in a localized part of the body constituted by discrete atoms. Also, the spatial relations of these atoms in the interstices and pores of the body within which they are arranged defined the separate ontological status and character of a given disease as distinct from the patient.[19] Hence the disease may even be construed as an invading entity existing in some sense on its *own* — thus having its own ontological status.

PLATO AND ARISTOTLE

Having identified the main features of the physiological and ontological disease models associated with the humoral and atomistic theories respectively, I would like to turn very briefly to the way in which health and disease is interpreted by Plato (c. 429–347 B.C.) and Aristotle (384–322 B.C.).

My task is complicated by the fact that neither Plato nor Aristotle supplies in his writings what one would call a full-blown theory of health. Rather, in both cases we are typically left with characterizations of healthy or diseased persons under various conditions. The more theoretically exacting premises are either tersely sketched or omitted altogether.[20] Yet it would be a mistake to conclude from this absence of detail a lack of interest in health matters. One need only recall the fact that most educated Greeks were quite abreast of the current theories of medicine. Indeed, as I have argued, both Plato and Aristotle took this background knowledge for granted when they borrowed from the medical arts whatever analogies and anecdotes served their instructional ends.

That Plato's view of health is marked by Pythagorean and Empedoclean influences, very likely absorbed from Philistion during Plato's stay in Sicily (at Syracuse), has been urged by A. E. Taylor and seems plausible enough. The Pythagorean emphasis on maintaining equilibrium, also present in the first principle of the humoral theory discussed above, is unmistakable, for example, in Plato's writings at *Republic* 444d 3–6:

But to produce *health* is to establish the elements in a body in the natural relation of dominating and being dominated by one another, while to cause *disease* is to bring it about that one rules or is ruled by the other contrary to nature.[21]

What is the significance of this passage? In the context of the *Republic*, where the discussion in Book Four turns on the parallel between justice in the state and justice in the individual, this passage serves mainly to establish a crucial analogy between the well-ordered body which is often called healthy, and the well-ordered soul which Plato calls just. That is, as health is the virtue (*arete*) or excellence of the body promoting its wholesomeness, beauty, and strength, so too justice is the virtue of the soul that fosters " . . . a kind of health and beauty and good condition of the soul . . . "[22] Thus the defect of the soul is the opposite of justice, namely injustice. What's more, injustice is pictured as a defect analogous to disease. So to ask, which is more profitable, being just or unjust, is to invite the conclusion that justice *alone* is more profitable. For to choose to be just is like choosing health; to become unjust is like courting disease. Hence it is fair to conclude that for Plato achieving a just soul is a necessary condition and, perhaps, even a sufficient condition for achieving mental health.

In addition, physical health is characterized by Plato as a correct balance between the elements of the body. (In his *Timaeus*, these elements are ultimately reducible to elementary scalene and isosceles triangles, and the four Empedoclean elements into which they are combined are given a broader explanatory role than the four humors.) Mental health is also characterized in a parallel way as a correct balance. Also at *Republic*, Book Four, the reader is informed that the three parts of a person's soul, reason, the spirited element, and the appetitive element, must be brought into proper attunement.[23] This attunement or balance is defined as manifest in a soul whose reason, allied with the spirited element (akin to emotions), benevolently governs the appetites (akin to bodily desires). Further, the three parts of the soul must work in harmony so that each is permitted to carry out *its own function* well without interfering with each other. In other words, it is supposed that reason has as its natural function the job of restraining the appetites; if the appetites usurped the authority of reason, some sort of disorder of the soul would ensue.

However, one may wish to challenge Plato's assumption that a theory of health, whether mental or physical health, can be formulated by simply reading off the putative natural functions attendant upon each part of the soul or organ of the body. It is fair to ask how Plato can know that the natural function of reason is to rule the appetites, and not vice versa. What evidence does he provide in this regard? Ultimately, he registers his own intuitions which are determined in the *Republic* by considerations of a normative sort regarding what form of government and what type of citizen would combine into the best possible commonwealth. But one can easily imagine another person thoughtfully submitting to his intelligence these same utopian considerations and getting a different list of "natural functions" and relations between elements of the soul or body. If so, this tends to cast doubt on there being a single, purely descriptive natural function for man, as Plato presupposes. Therefore, the starting point in Plato's theory of health, i.e., the assumption that one can discover man's natural functions, is flawed by oversimplification.

Whatever else might be said about Plato's views on health and disease, it is important to note that on his view the responsibility for carrying out a program of mental and physical hygiene rests squarely with each person. It is not the responsibility of one's physician, though the physician may make helpful suggestions. Thus, although some mental diseases in particular may have their contributing causes in (1) a defective bodily constitution inherited from one's parents, or (2) a bad upbringing, a person still has the duty to pursue those endeavors that will keep him well or help him overcome either

Plato believed its our moral obligation to cure our diseases (regardless of how we get 'em)

of the preceding two defects.[24] In other words, though one cannot fairly be blamed on Plato's view for either a defective nature or a defective nurture, one can be held accountable for remedying whatever defects of one's birth or childhood come to one's attention as an adult citizen. This is a moral obligation.

Although Plato's matchless student Aristotle went beyond his teacher by systematically defining the natural and biological sciences and pioneering the fields of comparative anatomy and physiology, there is little to be found in his writings on the explicit causes of human health and disease. Although his own father was a physician and there is every reason to believe that Aristotle was familiar with most of the medical theories of the Hippocratic Collection, not a scrap of evidence suggests that he himself practiced as a physician. But as one who directly laid the foundations of Western biology and, through Galen, indirectly influenced the course of medical thinking at least to the Renaissance, that portion of his writing that has been preserved on the topic of health deserves close attention.

One important notion that Aristotle borrowed from Plato's *Timaeus* was the conviction that nature is like an intelligent force that does nothing without a purpose. This teleological orientation looms large in the *Timaeus*, especially where it is asserted that the divine Craftsman who fashioned man was unenvious of his creatures and so aimed at what was best in creating the soul and body of man, their respective organs, activities, and functions.[25] Though as an empiricist Aristotle sought to wed his biological theories more closely to the observational data that he took pains to collect, both philosophers embraced the view (just criticized above) that at every level man manifests natural functions which may be determined by trying to discover the essential purposes of nature in designing a thing as it is.[26]

Moreover, Aristotle fell under the influence of the Hippocratic authors to the extent that he adopted their general characterization of disease as the disequilibrium of bodily structures or parts. He also allowed that seasonal or environmental factors affected man's disposition toward certain diseases. In addition, he associated the innate heat of the body with vital nutritional processes and rejected, as did the Hippocratics, any supernatural explanations of disease.[27] Yet in his writings he does not rely on their doctrine of the four humors. Instead, he identified the four physical opposites, the hot, the cold, the moist, and the dry, as the true elements of the natural world.[28] When these dynamic elements are properly blended in the human body, health results; when they are not, disease. And so in the *Physics* Aristotle states:

Thus bodily excellences such as health and a good state of body we regard as consisting in a blending of hot and cold elements within the body in due proportion, in relation to one another *or to the surrounding atmosphere*: and in like manner we regard beauty, strength, and all the other bodily excellences and defects. Each of them exists in virtue of a particular relation and puts that which possesses it in a good or bad condition with regard to its proper affectations, where by 'proper' affectations I mean those influences that from *the natural constitution* of a thing tend to promote or destroy its existence.[29]

Thus Aristotle, like the author of the Hippocratic treatise *Regimen*, generalizes that an animal's physical blend (as constituted by these four opposing elements) determines not only that animal's physical health but also its disposition (temperament), sensitivity, and intelligence.[30] Thus, the physical as well as the psychological characteristics of a given species and the individuals within that species are significantly determined by the unique blending of Aristotle's four elements.

At bottom, I conjecture that these four elements constituted for Aristotle the primary *material cause* of health and disease (both mental and physical). Though Aristotle does not say so explicitly, I further conjecture that he conceived the main *efficient cause* of a disease as lying in (1) the corrupting processes peculiar to the disease with which one is affected *and* (2) the choices one makes through adjustments in diet, exercise, and habits that may modify the course of the disease for better or worse. If so, both Aristotle's *formal* and *final causes* of disease can be related to the defective form of the affected organ or tissues of the diseased individual. In this sense, some diseases may be understood in Aristotelian terms as the privation or corruption of the relevant natural function properly associated with the form and purpose of the animal part in question.[31]

Given the above interpretation, which is largely an effect of reconstruction on my part, an additional implication of Aristotle's theory of health may now be explored. Doubtless Aristotle wanted to resist a wholly deterministic theory of disease so that some individual responsibility could be assigned for one's own health. The individual's habits and way of life constituted part of a regimen that the individual was free to adjust in order to suit his needs, aims, and values. In other words, the individual was free to choose what foods to eat, what drugs to take, what climate to reside in, how much exercise to take, etc., to the point where he could fairly be held at least partly accountable for the state of his own health. In this sense, the agent himself had the capacity to be part of the efficient cause of his own disease and its cure.

This fact of personal accountability, in my opinion, is what Aristotle was alluding to when he said: "Old age is a natural disease, while disease is an

acquired old age."[32] That is, old age (with its normally accompanying infirmities) is the natural outcome of a healthy adulthood. No one can be blamed
for growing old. But disease acquired prior to old age may in some cases, at
least, be viewed as having something to do with the way an individual has
chosen to live. This result is also quite compatible with Aristotle's ethical
teachings. He taught that with few exceptions a person may be held accountable for his own welfare, i.e., the state of his being.

THE MEDICAL SECTS

Following the deaths of Hippocrates and Aristotle in 399 and 322 B.C.
respectively, almost all of the medical literature of the ensuing 400 years
has been lost to posterity. But such theories as there were during this period
up to the birth of Galen in 129 A.D. have been reconstructed from fragmentary remains and from the illuminating commentaries of Galen and the
post-Galenic authors. What can be said with some certainty about this intermediate period between the death of Hippocrates and the rise of Galen is
that philosophical differences with regard to the way medicine should be
practiced intensified among the Greek physicians to the point where four
distinct medical sects gradually emerged. Listed in rough chronological order
these were: the Dogmatists, Empiricists, Methodists, and Pneumatists.

As I have already suggested, there never was any such thing as a primitive
Hippocratic doctrine in the first place. In other words, the Hippocratic Corpus
itself emphasizes a number of competing medical treatments, procedures,
and theoretical assumptions. Perhaps owing in part to the general emergence
of individualism associated with the Hellenistic period and the disintegration
of the Greek city-states, as well as to the increasingly fashionable philosophical partisanship evidenced in the fragments of numerous post-Hippocratic
Greek physicians, this original divergence of opinion with respect to both
the theory and practice of medicine became more vocal. As to the different
philosophical tendencies and treatment strategies represented by the four
sects under discussion, scholars have so far reached no clear consensus on the
detailed beliefs associated with each sect. Nor would it serve the basic aims
of this inquiry to set forth here their conflicting assessments.

But in general it can be usefully noted that the Dogmatists emphasized
inferring the hidden causes of disease from manifest symptoms. They were
greatly influenced by the humoral theory of the Hippocratic authors as well
as by the teleological orientation of Plato's *Timaeus*, which they quoted
chapter and verse.[33]

The Empiricists reacted against what they felt was the Dogmatist obsession with overblown theories founded on an absurdly small number of observations. They emphasized instead closer observation of each individual patient and removal of the disease symptoms without necessarily trying to figure out the hidden causes of disease. They were most skeptical about being able to detect hidden causes. The Empiricists were influenced both by the skepticism of Pyrrho (fl. 340 B.C.) and, ironically, by the empirical orientation of Aristotle — though it appears they never fully appreciated the latter's respect for finding out the causes of phenomena.

The third major sect was the Methodists. They may be regarded as somewhat *intermediate* between the theoretical tendencies of the Dogmatists to detect hidden causes for diseases in the absence of empirical research on the one hand, and the more practical tendencies of the Empiricists to cure the patient of his suffering in the absence of ascertaining the ultimate causes of diseases on the other. The Methodists were influenced by the atomic theories of Democritus and Epicurus. They sought to understand and cure disease based on the theory that extremes of contraction or relaxation in the pores (areas of empty void) of the body caused all illness.

In contrast, the fourth major sect, the Pneumatists, were associated with the Stoics. But they were also frankly eclectic in their approach to healing. They were the strongest opponents of the Methodists in Rome, whose theory they regarded as oversimplified. Alternatively, they postulated that health and disease were determined by the condition of the pneuma, or "vital air" each person breathed. But the condition of one's pneuma (which could be monitored in the pulse) was said to be caused in turn by the imbalance of the four qualities hot, cold, moist and dry, plus the four humors they associated with these qualities.[34]

Finally, whatever the merits of the medical theories sponsored by the preceding four sects may have been, it is fair to conclude that the absence of any one dominant theory was due largely to the related absence of any acceptable criterion for theory selection among the various members of the medical craft. So the partisans of each sect went their separate ways largely ignoring the theoretical objections of their opponents.

THE GALENIC CONSTITUTIONAL THEORY

Turning to the writings of Galen (129–199 A.D.) on the issue of health and disease, one is at first struck by the wide range of his contribution to medicine and philosophy in general. Benjamin Gordon states that eighty-three of

Galen's works are still extant and are acknowledged to be genuine. These include fifteen books on anatomy alone, plus treatises dealing with logic, ethics, and grammar.[35] Given the sheer scope and output of the writings of this last prominent Greek physician, the most that I shall undertake in the remaining pages of this chapter is to critically discuss the eclectic nature and therapeutic implications of Galen's thought. This I shall do by highlighting some of the ideas he borrowed and refined from the previous thinkers we have encountered.

Galen was a follower of Hippocrates. All of the seven medical principles associated with the Hippocratic Corpus were known to him and were by and large accepted with various modifications. From Plato and Aristotle he revived the religious and teleological notions that, in Galen's own words, "There is nothing in the body useless or inactive, but all parts are arranged to perform their offices together and have been endowed by the creator with specific powers." [36]

Moreover, Galen's method was essentially empirical and systematic: he sought to reduce all knowledge to general, interrelated principles based on observation. At the same time, he resisted following any one medical sect or school. He regarded those fellow physicians who did as slaves. Yet, notwithstanding his original contributions to experimental physiology and human anatomy (he proved that *both* arteries and veins carried blood and correctly identified for the first time many of the neurological functions of the brain and spinal cord), it must be admitted that many of Galen's anatomic concepts were flawed by his habit of reasoning by analogy. That is, he assumed that what was true of the anatomical structure of animals would be equally true of man. Galen also possessed an over-zealous tendency to read into all of the structures and processes of the body the purposefulness of nature. Even so, Galen's studies of human anatomy were the best among the ancients and remained so right down to the time of Vesalius in the sixteenth century.

More specifically, Galen's writings on the causes of disease reveal a creative attempt to synthesize the physiological concept of disease and the ontological concept of disease discussed earlier. That is, I shall argue that Galen tried to explain disease based on the natural functions of the body as well as on the determinate structure of the body's composite parts.

The physiological components of his theory Galen borrowed from three basic sources: (1) from the various humoral doctrines of the Hippocratic and Dogmatic authors; (2) from the four Empedoclean elements and four physical opposites adopted by Plato and Aristotle respectively; and (3) from the emphasis placed on the vital breath developed by the Pneumatist authors. These antecedents are evidenced in Galen's physiological orientation by the fact that he took the Aristotelian opposites as basic to the organizational hierarchy of the body. In addition, just above these Aristotelian opposites he postulated the Empedoclean elements. These were followed by the humors (which Aristotle had little use for). Thus it may be said that more distinctly than Aristotle or any earlier Greek thinker, Galen emphasized the general integration (*systasis*) of all the parts of the body to compose a functional whole.[37]

In addition, pneuma held an important place in Galenic pathology. He believed that the function of the nervous system was to carry the pneuma — which he also identified as the life-giving principle — to all the parts of the body. Moreover, like the humors, the pneuma was regarded by Galen as a physical and not merely a metaphysical substance. He did not regard pneuma as a fixed entity blown into the body at birth or conception. Rather, it was something that needed to be constantly renewed through the respiration of the inspired air of the world soul. (The Stoic influence just here is obvious.) Also, he theorized that the inhaled pneuma became the *animal spirit* when it came in contact with the brain and nerves; the *vital spirit* when it reached the heart and arteries; and the *natural spirit* when it reached the liver and renal veins. Hence in this regard he devised something like Plato's three-part division of the soul.[38]

But in other respects his psychology resembles Aristotle's. For he seems in the final analysis to count the body and soul as inseparable parts of a single integrated system. In addition, unlike Aristotle, but like Alcmaeon, Democritus, Hippocrates, and Plato, Galen links the brain with the center of thought. Of his three postulated spirits, moreover, he says that the animal spirit lodged in the brain is the noblest. To the animal spirit, which he associates also with the reasoning faculty, he ascribes the Aristotelian efficient, formal, and final causes of the soul's motions, perceptions, and sensations. He claims that a person's perceptions and sensations are communicated to the rest of the body through the spinal cord and its network of nerves.

Thus, for Galen health and disease can be understood on the physiological level partly in terms of the condition and natural function of the patient's pneuma. Following the Pneumatists, Galen also implies that there is some sort of soul-body causal interaction. For he allows that the undifferentiated pneuma *originates in the blood* through the circulating action of the heart's innate heat and the lung's respiration. Hence in this brief look at the physiological side of Galen's eclectic theory of health, we also come to appreciate his psychosomatic justification for bloodletting which he sometimes prescribed for both physical and mental disorders (e.g., melancholy). For if the blood was suspected of being impure or defective in some way, the logic was that perhaps one's psychological condition could be altered for the better by ridding the body of any such imbalance or impurity contained in the blood. What's more, it comes as little surprise that from this partial causal dependence of mental states and processes on the bodily condition of the blood (not to mention the ascending hierarchy of opposites, elements, and humors), Galen ascribes such psychological disturbances as mania, imbecility, and dementia to somatic malfunctions of various sorts.

But Galen is not content to ascribe the processes of disease and health to the physiological functions of the body alone. The explanatory burden of Galen's system is shared by a structural analysis that focuses on the arrangement (*diaplasis*), size, and number of each of the body's constituents as well. This approach resembles what I have earlier called the ontological concept of disease. Moreover, here Galen appears to be somewhat influenced by the early Cnidian tradition, and, if not by the atomists Democritus and Epicurus, then certainly by the Methodist sect that was very active in the Rome of his day.

Yet it must be acknowledged that at times [39] Galen wrote that health entails a due proportion or blend of the four physical opposites hot, cold, dry, and moist. This sounds like the physiological model alone. But in his treatise *On the Best Constitution of Our Body*, Galen elaborates on his theory of health in a crucial way. He clarifies the point that the proper blend of the four opposites is merely a necessary and not a sufficient condition for health. In other words, the proper blend of the four opposites is at most an essential ingredient of health; it is no guarantee of health taken alone.

In addition, he says that the constitution of the body is determined not only by the blending of the four opposite qualities in each of the body's like-parted or homogeneous parts (e.g., tissue, blood, bones). Also, the constitution of the body is determined by the higher *arrangement* of these homogeneous parts into heterogeneous parts comparable to those we call organs (e.g., hands, eyes, etc.).[40] The upshot of his message is that proper blend of the opposite qualities *and* proper arrangement of the body's parts are the individually necessary *and jointly sufficient conditions for health* in the body with the best constitution.

This preceding scheme led Galen to assert that the best constitution of the body is one that most efficiently resists *external* diseases due to such things as climate, excessive exercise, etc., and *internal* diseases due to such things as the "residues of food."[41] How, then, can one achieve the best bodily constitution? The answer is that it can be achieved just in case the two criteria of proper blending of the four opposites *and* proper arrangement (*diaplasis*) of the body's parts are met. The medical result of these twin criteria is that disease can be analyzed on two levels: (1) temperamentally (in terms of blending) or (2) diaplastically. Galen refers to both of these criteria simultaneously in the following passage. In speaking of the dense and rare, he is referring to the size, shape, and arrangement of the pores in the heterogeneous parts of the body as well as to the opposites, dry and moist. When these pores were open and moist, he termed them "rare"; when they were closed and dry, "dense." [42]

Thus, we can say that the well balanced body, in addition to its other advantages, will be neither rare nor dense. It will be midway between these extremes, just as in other respects; it has the advantage in some way over each of the two excesses. For the denser body is less subject to external causes of disease, and the rarer less subject to internal ones. And though you could not find a body that is strictly resistant to both causes, you might say that the body that is midway between all extremes is resistant to both *in a relative sense* − and we assert that such a body is the healthiest of all . . . . Thus, . . . the mean is not the most completely resistant of all conditions; but though it may be inferior to every condition in some particular way, still it is the most desirable of all. [43]

Hence, an optimally constituted body Galen formally defines as one that manifests *an intermediate position* between various excesses and deficiencies with respect to both the temperament (blending) and the arrangement (*diaplasis*) of its parts. This is so because only when these twin conditions are met (and Galen outlines the specific excesses and deficiencies in considerable detail) will the body resist disease that otherwise brings disruptions upon the body's normal functions. [44]

In sum, these disruptions can be viewed in a manner akin to the ontological disease model as having a specific location in the structure of the body. In that case removal of the diseased part by surgery or purgation may be undertaken. Or, *more typically* for Galen, these disruptions may be viewed on the physiological disease model. In that case, adjustments in diet, exercise, or the application of drugs may correct the malfunction. It is necessary to add, however, that true to his eclectic orientation not one but many etiological schemes are explored by Galen. Therefore one must be careful not to oversimplify by casting all his ideas about health and disease in the single form just discussed.

A final point. In the preceding long quotation as in his other writings, at least two tendencies emerge. As each of these two tendencies have important therapeutic implications, I would like to digress for a moment and explore what they are.

First, from the fact that each person is regarded as having a unique temperament and a unique arrangement of bodily parts, it follows that no two people are exactly alike. This theoretical conclusion would explain the well-known fact that, given two patients with the same disease, a remedy that will work on one patient may not always work on the other. Therapeutically, these considerations led Galen to emphasize that the physician must get to know the individual characteristics of each patient *before* deciding on treatment no matter how familiar he may be with the patient's disease. (Just here Galen would agree with the Empiricist sect on the importance of closely observing the patient's traits; in general, though, he found the Empiricists' disdain for identifying the internal causes of disease totally repugnant.)

Second, in the preceding quotation Galen also appears quite willing to define health "in a relative sense." That is, health is an intermediate condition between opposing extremes. He tacitly admits that the best one can hope for is a bodily constitution that is not quite perfect since, in fact, "it may be inferior to every condition in some particular way." Furthermore, in the aforementioned treatise by Galen there is a passage in which he states: "For *absolutely perfect* functioning in every respect, so that none of the homogeneous parts and none of the organic parts is imbalanced, is not something that tends to occur very often, though *instances* may occur over longer stretches of time." He also states: "We have shown in our discussions of it that health is not something narrow and absolutely simple and indivisible, but rather capable of wide variation." [45]

What these passages point to, I am persuaded, is the relaxation of the earlier Greek Archaic and Hellenic notion of health as an absolute ideal (as Plato, for example, tended to view it in his *Republic* and *Timaeus*). Galen here moves toward a more variegated or "second best" notion of health adjusted to meet the exigencies of daily life as it was led by the average citizen. That is, only the rich and idle were in the position to follow the strict regimen of proper exercise, climate, diet, rest, etc. that was typically prescribed by the Greek physicians (especially those of the Classical period). By the Hellenistic and later Roman period, economic necessity and the political pressures of being an active, involved middle-class citizen trying to earn a decent living may have combined to force *some* of the Greeks and Romans to modify their old belief in the existence of perfect health and the desirability of attaining it.

Addressing this point, Fridolf Kudlien has plausibly suggested that another factor in the relaxation of the ideal of absolute health may have been the intellectual movement associated with Protagoras' ethical relativism and, I would add, later skepticism.[46] Edelstein, moreover, has traced the trend to settle for health in the relative sense to the Hellenistic period's increasing tendency to recast health in terms of the individual's own choices. These choices are of course *relative to* an individual's felt needs and self-knowledge. If so, perhaps the notion of health as a delicate balance requiring the coddling of the individual outlasted its usefulness as more and more individuals increasingly sought to take charge of their own regimens. No wonder, then, that the Roman period evidences a fair amount of resentment and ridicule directed at the fawning counsel sometimes given by doctors.[47]

Whatever the specific pressures of change may have been, there is little doubt in my mind that more than simply Aristotelian language is at work in Galen's talk of the best bodily constitution in relative, not absolute terms — terms also reminiscent of Aristotle's ethical theory of the mean (virtue).

Therapeutically, the consequences of the idea of relative health seem to be helpful. On the one hand, the average patient settles for the fact that at any given time, though he may *feel* no pain or discomfort and though he may consider himself relatively healthy, he may at the same time be constitutionally disposed to get one sort of disease rather than another. If so, the physician ought to prescribe a regimen relative to the patient's own constitutional needs. The therapeutic aim would be to hold off those diseases to which the patient is more or less disposed.

On the other hand, those patients who are not able-bodied because they suffer some defect like loss of a limb may yet be usefully regarded as having "sufficient health" relative to their situation. This would seem justified provided such patients are judged free of all other disease and are somehow able to cope with their defect.[48] Thus the amputee may have been regarded as healthy in the relative sense. This way of looking at health may have lessened somewhat the stigma especially associated with such losses in antiquity when fewer restorative measures were available.

Lastly, the notion of relative health doubtless cast the doctor in a more realistically limited role: the doctor could perhaps at least improve *if not totally restore* health. And for the ancients, as for us, such relative improvement in and of itself surely has value. This, too, Galen seemed to appreciate more than did many of his predecessors.

CHAPTER THREE

ATTITUDES TOWARD DEATH

No matter how much the Greeks valued health, they realized that sooner or later even the healthiest and most robust among them must die. It is my purpose in this chapter to illustrate some of the attitudes that the Greeks had toward death at various times in their cultural development. What I shall argue is that the Greeks had no single view on the meaning of death. Rather they held a plurality of often conflicting views, just as we do in our own culture today.

Before undertaking to sketch their various attitudes on death, it is imperative to isolate precisely what features of this complex topic one seeks to understand. The simple expression "attitudes toward death" involves a complex of interrelated concepts that are by no means identical.[1] Attitudes may be explored on at least three levels: (a) cognitively, one may inquire how people think about death; (b) affectively, one may try to discover how people feel about death; and (c) behavioristically, one may investigate how people behave toward their own or others' impending death. My analysis will encompass primarily (a), the cognitive side of this topic. I shall ask, How did various philosophers, poets, artists, and ordinary Greek citizens conceive of death at different points in Greek culture?

Furthermore, the very word "death" is itself a highly elusive and disunified concept in our ordinary usage of that term. Its standard senses in the present context include: (1) the dying process; (2) the precise moment of death; and (3) the state of being dead. I shall be primarily interested in (3), how the Greeks conceived of the state of being dead. The answer to this question promises to significantly contribute to our background understanding of various Greek ethical views on euthanasia, infanticide, and suicide. Yet (1) and (2) are also important to this inquiry and so will be treated here as subsidiary topics.

Moreover, for the purpose of further clarifying our focus it will be useful to distinguish between the natural processes of aging and the actual dying process in the last days of a person's life. Otherwise one might get ensnared into the confusion of saying that each person is dying from the moment of birth, which is at best a sensational but not very accurate appraisal. Instead, it will be assumed that the dying process entails having reached a critical and life-threatening stage in the aging process which, if left unchecked, would normally be expected to result in death. So it is not Greek views on merely aging or growing old that I shall address. Rather, it is how the Greeks approached dying, as just defined, that will be of special interest.

Lastly I must say a word about the inherent limitations of the evidence and method on which the following sketch rests. Basically, I am confining my discussion to the treatises, poems, and fragments of selected philosophers and poets plus secondary sources judged to be of value. Although some effort will be made to balance the main strands of thought, surely many other sources could be fruitfully included. In what follows I shall mention just a few of the Pre-Socratics; and the Stoics for the moment are passed over in favor of the Epicureans. Even so, within these stated limitations a representative picture of Greek death may be secured.

DIVINE AND CHTHONIC PERSONAL IMMORTALITY

If one turns to compare the depictions of the state of being dead as evidenced on the one hand by the epic poetry of the Homeric religion (850 B.C.?), and on the other hand by the Orphic tendencies manifest in the religious doctrines of the Pythagorean Brotherhood (c. 500 B.C.), at least one major difference in outlook comes into view. The Homeric religion generally depicts death as unconquerable and terrible. The dead are shown as bloodless shadows wandering listlessly in the underworld.[2]

In contrast, the Orphic-Pythagorean religion depicts death as conquerable and promising, and claims that the best and most essential part of each person, his soul, is deathless and divine. It teaches that through purification the divine soul is destined finally to reunite with the divine cosmos from whence it originally came before becoming entombed in the inferior mortal body, which alone is subject to decay.[3]

This preceding divergence in outlook can be further characterized by calling the dominant Homeric perspective on the state of being dead *chthonic personal immortality*. By using the word "chthonic" (taken from the Greek *chthonios* meaning of or in the earth), I wish to emphasize the fact that the Homeric dead are essentially spirits of the underworld, mere shadows of their former selves.

The Orphic-Pythagorean perspective, on the other hand, I shall call *divine personal immortality*. The decisive difference between these two perspectives, it seems to me, turns on the way each characterizes (a) the initial relation of the soul to the body; and (b) the further relation in which that part of each person that survives the moment of death (often called the soul) stands to the enduring order and structure of the universe.

As to (a) above, the initial relation of the soul to the body, Homer does not grant the soul of man divine status. Nor does he oppose the worth of the soul to the comparative worthlessness of the body. It would, however, be strictly speaking wrong to say that he makes no distinction between the two. For Homer (as for Hesiod), *psyche* has two meanings. It means both "the breath of life" and "the individualized ghost." Typically, the ghost or "shade" escapes (like animating breath) from the mouth, wound, or limbs of the dying hero at the moment of death. It leaves behind the *soma*, which for Homer literally meant "corpse." [4] Then the ghost repairs to the underworld to endure a benign, joyless existence. Hence it may be argued that there is some hint of a soul-body (or psyche-soma) dualism in Homer but only in the narrow sense that he allows some quite diminished piece of each former person, namely a ghost, to survive the body at death. After this separation of ghost and corpse, the ghost mourns for its former body as one might mourn the loss of a worthy friend.

As to (b) above, the relation of the ghost to the ongoing universe, with few exceptions a former person's surviving ghost stands in a decidedly miserable and lowly relation to the rest of pulsating humanity left behind, as well as to the only truly deathless ones — the gods. That is, the Homeric ghosts live a shadowy existence in a dim and dank underworld which the gods themselves can't stand. We get some idea of the undesirable status of the dead from the *Odyssey*, Book II. There the ghost of Achilles confronts and scolds the full-bodied Odysseus, saying:

O shining Odysseus, never try to console me for dying. I would rather follow the plow as a thrall to another man — one with no land allotted him and not much to live on — than be a king over all the perished dead. [5]

Here the judgment of Francis Cornford in his book *From Religion to Philosophy* seems entirely borne out. Cornford has it that in the Homeric religion there existed a definite gulf between the divine and the merely human. [6] In this connection, I am inclined to think that the Delphic motto "Know Thyself," which admits of many interesting interpretations, may originally have served as a reminder to mankind that this gulf was real and had to be respected at all times. It was hubris to ignore this line of demarcation and begin to think of oneself as a god.

Pythagoreanism again stands in stark contrast to this Homeric picture of man's relation to the universe. Waiving the sticky question of the origins of

the three Pythagorean doctrines of (1) the essential divinity of the soul,
(2) the possible purification of the soul, and (3) the transmigration of the
soul, it is certain that the collective result of these three doctrines was to
remove or significantly reduce the felt gap between the human and the divine.
This is partly so because the Pythagorean soul is taken to be a fallen divinity.
The body serves as a temporary tomb from which it can win eventual release
through the study of philosophy and mathematics. Also, the soul can in this
way recapture its divine purity and be restored to its full status as a truly
deathless part of the harmonious cosmos.

Furthermore, I have suggested that, in contrast, such a sublime prospect
for the deceased is not central to Homeric thought. Moreover, it is a curious
fact that during the Archaic and Hellenic periods, at least, more Greeks by
far adopted variants of the Homeric chthonic view of death than the competing Pythagorean divine view. Why this was so, I shall leave for others to
speculate; doubtless the elitist nature of Pythagoreanism was a chief factor.

In any case, before turning to a third perspective on the prospect of survival beyond death, a brief digression to explore a few features of the grammar
of the Greek word for death, namely *thanatos*, is appropriate.

Emily Vermeule has discovered in her book *Aspects of Death in Early
Greek Art and Poetry* that the Greeks "had no word for irreversible death;
one does not die, one darkens," she suggests.[7] What sense can be made of
this? Vermeule thinks that it signifies a basic hope that the Archaic period
Greeks (c. 700–478 B.C.), at least, held for the imperishability of the *psyche*.
"The *psyche*," she says, "is memory and intelligence dormant and can under
proper circumstances be recalled [to life] by the grief, love, magic or poetry
. . ." of the living.[8]

But this hope, however strongly felt by the Archaic period Greeks, must
surely have been sensed to be in vain, even for the followers of Homer. Sooner
or later almost all ghosts will have been forgotten. Inevitably, the community's
memories of the deceased fade from the collective consciousness of succeeding
generations of mortals. The "silent majority," who after all constitute the
Homeric dead, are destined – with the exception of a few remarkable heroes
of old – to be recalled never again by either the eulogies or encomia of the
living. Little wonder, then, that one of the favorite artistic renderings given
the dead person's ghost by Archaic vase painters in the seventh and later
centuries, was to depict the ghost as a tiny, insignificant butterfly or soul-
bird. Typically, on funeral urns the soul-bird was shown flitting about in the
afterworld in assured anonymity.[9] It could almost be swatted by the living
as if a bothersome mosquito.

What's more, in addition to being a common noun, *Thanatos* is also a proper noun designating the personage of Death, who is the twin brother of Sleep (*Hypnos*) in the Homeric pantheon. As may be seen from the vase painting on at least one magnificently large red-figured krater dating from 515 B.C., and painted by Euphronios (see frontispiece), Thanatos is not pictured as a devouring or terrible figure. Rather, he is shown more like a resolute and avuncular escort: someone with whom one cannot argue. As he is in the present specimen, Thanatos is often rendered as a large, winged male figure cloaked in battle gear. He has piercing eyes and somewhat over-sized hands and feet. On the vase just mentioned, Thanatos is shown lifting the body of the Lycian prince Sarpedon at Troy, with his brother Hypnos assisting. Elsewhere, he may be symbolized poetically as a black cloud, or as a hue or mist appearing around the head of one about to die.

Moreover, Thanatos, the personage of death, is not a fully developed figure in the Homeric epics. He appears in the flesh only once in Homer, at *Iliad* XVI. 671. There, the aforementioned Sarpedon is dead, stripped, looted, and surrounded by enemies. He is rescued by Thanatos — who cradles him like a fellow kinsman. The twin brothers, Sleep and Death, then escort the corpse to Sarpedon's native Lycia for proper burial. Incidentally, Thanatos is almost never characterized as a killer or as an agent of death. Nor is his ally Hades, the dreaded king of the underworld.[10] Perhaps this may be explained in part by the fact that many Greeks regarded death as the negation of life, as a cessation or an inversion of living.

NATURAL PERSONAL DISSOLUTION

If there is yet a third Greek perspective on the state of being dead, Jacques Choron may be on the right track in looking for it among the early Pre-Socratic philosphers of Ionia. In his book *Death and Western Thought*, Choron argues that the favored irreducible elements of the earliest monistic philosophers such as Thales' water, Anaximander's boundless (*apeiron*), and Anaximenes' air (*aer* or *pneuma*), signify an effort on their part "to neutralize death." What's more, he suggests that it "may be an expression of the hope that somewhere and somehow death shall have no dominion."[11] I shall argue that Choron's interpretation with respect to these three monists is interesting but, as it stands, paradoxical. My focus will be on the lone fragment of Anaximenes, but what I have to say about Anaximenes can apply as well to Thales and Anaximander.

The only authentic fragment attributed to Anaximenes (fl. 546 B.C.) translates:

As our soul, being air [*aer*], holds us together, so do breath [*pneuma*] and air surround the whole universe.[12]

Most scholars agree that for Anaximenes air is the cause of all things, for it "holds us together" and presumably controls both man and the cosmos. Further, air is the basic constituent out of which all things are born and into which they shall return. In identifying man's soul with air, Anaximenes seems at the same time to be linking air to the principle of life. It is likely that for him air was the necessary condition for life in all living creatures.

Furthermore, like Anaximander's boundless and Thales' water, Anaximenes took air to be deathless and ageless: it is the only thing that endures through change. Though it is true that he ascribes an airy quality to the soul, nothing Anaximenes or other ancient commentators say attributes to him the further crucial identification of the soul as the real, true, or essential person. This identification, though it is sometimes neglected, is crucial to any interesting notion of personal survival or immortality. That is, it serves as the necessary condition for that sort of personal immortality that links the survival of the soul with survival of the continued psychological identity of the deceased. Anthony Flew has underscored the importance of this crucial link in graphic terms. "Unless it is established that I am my soul," he writes, "the demonstration of the survival of my soul will not demonstrate *my* survival, and the news that my soul will last forever could provide me with no more justification for harboring 'these logically unique expectations' than the rather less elevated assurance that my appendix is to be preserved eternally in a bottle."[13]

Now the crux of Choron's interpretation of the early monists is this. If all things are one, as they claim, then the most shocking change of all is going from life to death, But, Choron reasons, death itself becomes less radical owing to the essential oneness of all that exists.[14] If by "less radical" he is referring to the ultimate reducibility of all phenomena (including death) to a single, underlying principle (water, the boundless, air), he may have a point. But he also implies that somehow the unified, monistic scheme provides some sort of *psychological* consolation. This it allegedly does by reducing the horror of the radical physical change each person will undergo when at death one begins to lose the unique individuality on which one's personal identity and continuity may at least partly depend. If so, was the specter of death thereby "neutralized" or made to seem "less radical" for these three Pre-Socratics? It is not easy to speculate on such matters. But Choron's guess that just the thought of "the everlastingness of the Whole" somehow neutralized death finds no backing in the relevant fragments of Anaximenes, Anaximander, and Thales. Personally, I think that Choron is delivering too modern a reading of these early monists' views on death. It is just as likely that they found the finality of death, with the *prima facie* elimination of personal survival, as hapless and as perplexing a prospect as did many of their less gifted contemporaries.

Choron's analysis of the early monists is further damaged by his use of the same word, "death," in two distinct senses without being aware of it, thereby committing the fallacy of equivocation. This happens when he states that these Pre-Socratics, "accepting their individual deaths, pin their hopes on the everlastingness of the whole, in which [case] death cannot be senseless perishing." The first occurrence of the word death is used in the *strict sense* to mean the exclusion of personal survival, as when we say Smith met his death$_1$ in a tragic fall. But the second occurrence is used in the *loose sense* to leave open the possibility of personal survival, as when we say Jones believes in life after death$_2$.[15] That this equivocation is going on can be readily appreciated by noticing that unless the second occurrence of death leaves open the possibility of personal survival, there would be no enduring subject on which such "hopes" in the relevant sense could be "pinned."

Nevertheless, I must admit that despite these flaws, Choron's attempt to read the early monists as if they supplied a third perspective on the state of being dead is suggestive. Hence I shall call this third perspective *natural personal dissolution*. This label emphasizes their view that each unique person terminates at death, following which his constituents gradually dissolve and reform into various irreducible physical elements.

This perspective, with adjustments, could also characterize the basic outlook of several more Pre-Socratics. The notable *exceptions* are Heraclitus, who admits no fixed boundary between the processes of life and death, since everything is changing;[16] Parmenides, who holds so fast to the notion of the imperishability of Being that death would for him have to count as a mere sensory illusion;[17] and Empedocles, who in his poem "On Nature" seems to admit natural personal dissolution, only to contradict himself in his religious poem "Purifications" by talking as if he believed in divine personal immortality.[18] Finally, if this third view of death does succeed in somehow neutralizing death, perhaps the fundamental way it accomplishes this is by replacing the frightening visages of death sometimes associated with the chthonic perspective of Homer, with the seemingly more neutral notion of personal dissolution *unaccompanied* by any possible torment, loneliness or regret in some afterworld existence.

Presently, though, I would like to leave the Pre-Socratics to consider what some of the leading Greek poets and tragedians of the sixth through fourth centuries B.C. had to say about death. In addition, I intend to examine briefly some of the commonly held fifth century beliefs about the dead.

POETIC AND POPULAR IMAGES

According to the poet Sappho (fl. 582 B.C.), death is the greatest evil. "Gods so consider it," she says, "else they would die."[19] This impeccable logic finds company in the lyric poet Anacreon (fl. 540 B.C.) who is similarly horrified by approaching death: "Death is too terrible. Frightening are the depths of Hades. There is no return."[20]

If these words of Sappho and Anacreon express the customary pessimism connected with the Homeric picture of chthonic personal survival, a far more optimistic estimate (probably inspired by Orphic ideas) is expressed by Pindar (518–438 B.C.). According to this view, which is a variant of the Pythagorean's divine personal immortality, the soul lives on after death enjoying much brighter prospects. Pindar writes:

By happy dispensation all travel to an end which sets free from woe. And the body, indeed, of all goes with mighty Death. But there remaineth alive a phantom of life; for that alone cometh from the gods.[21]

The phantom to which Pindar refers is of course the divine soul of man. That this soul comes from the divine gods, as Pindar says, illustrates again the tendency of this particular perspective to bridge the gulf between the gods and man – a gulf so evident in the earlier Homeric poems.

Now, among the works of the three major Greek tragedians, Aeschylus, Sophocles, and Euripides, there appears on the whole to be an acute death consciousness. Thus the chorus in Sophocles' (496–406 B.C.) play *Antigone* declares: "Of all the great wonders, none is greater than man." Then, almost as an afterthought, the chorus laments: "Only for death can he find no cure."[22]

Yet Sophocles also allows us a glimpse of what might be called *the heroic attitude* toward death. According to C. M. Bowra, "the exponents of the heroic ideal regard death as the climax and completion of life, the last and most searching ideal to which a man is subject and the true test of his worth."[23] In my opinion, this heroic attitude was doubtless admired, but in actuality was probably as rare among the Greeks as true heroes were in real life. Antigone represents this heroic attitude toward death when she asks her self-righteous accusers: "Is there something more you want? or just my life?"[24] Antigone demonstrates that for the sake of preserving the high ideals of justice and honor, a worthy person should be prepared to meet death without the slightest compromise. This she did for the sake of insuring that her outcast brother, Polyneices, would received a proper burial as divine law, if not human law, required.

In contrast, there are some passages in Aeschylus (525/4–456 B.C.) that probably come closer to reflecting the general fifth century Greek attitude about death than either the Orphic or heroic visions just mentioned. It will be recalled that on the Homeric view of death there is little hope of a blessed immortality to comfort the dying or lessen the grief of surviving friends and relatives. A. W. Mair has argued that, when death is spoken of as desirable at all, it is pictured merely as a refuge from evil (*kakon kataphyge*), a dreamless sleep. This is the picture of death Aeschylus presents when his chorus bewails the murder of Agamemnon in his play devoted to that tragic figure. The chorus laments:

Would that some fate might come speedy, not over painful, nor with lingering bed, bringing to us the everlasting, endless sleep.[25]

But I wonder: is Mair overlooking that the Greeks also possessed the concept of euthanasia (εὐϑανασία)? Probably not. For "euthanasia" was used to mean a good, easy, or noble death, a death evidenced by a minimum of pain and distress. It was not a concept that carried with it a higher or friendlier vision of the afterworld. Fundamentally, euthanasia involved the context of dying, not the state of being dead.[26] The Greek concept of euthanasia, then, includes precisely that permanent escape or refuge from the evil of relentless pain to which Mair here alludes. It does not entail any particular perspective about the afterworld.

As for the related issue of how the living tended to perceive the dead in the Classical period, the primary evidence has been culled from the art, law, and literature of that age and reconstructed from the evidence found at ancient burial sites. What has crystallized from this evidence and what can be affirmed with some certainty may be summarized as follows:

(1) There was a tendency to believe that the dead could somehow perceive what is done and said in the world of the living.

(2) The presence of the dead was in some sense felt; their goodwill was valued, and their contempt feared.

(3) A person who was murdered was thought to have the potential of becoming an angry ghost capable of causing sterility in crops and beasts, as well as causing other fiendish calamities.

(4) The denial of a proper burial, irrespective of the cause of death, was thought to anger the soul of the dead man and also assured misfortunes for the living.[27]

Although I cannot pursue the vast implications of these four beliefs in detail, I would like to offer a few brief comments on some aspects which touch primarily on the issue of moral conscience and moral responsibility.

With respect to (1), it has not yet been positively established that belief in the capacity of the dead to perceive the affairs of the living had an affect on moral conduct. But even a wavering belief that such a possibility *might* exist may well have conditioned some of the moral reasoning and conduct of the living. At least, this possibility cannot be altogether discounted.

As to (2) – (4), it will be noticed that the alleged capacity of a bodiless ghost to influence causally the full-bodied living raises a hornet's nest of questions about just how such a causal connection would work. That is, how can a bodiless ghost impinge in any way on a thing which has body? The answer to this dilemma seems to be that it worked through some sort of supernatural agency, the full details of which were understood only by the gods. No systematic eschatology was ever worked out by the Greeks to explain away all the possible objections which might be prompted by considerations of logical consistency. Moreover, as K. J. Dover has pointed out, "The ghost's capacity for direct action on its own initiative and at its own discretion was regarded as limited; it was rather the gods' interests in the *rights* of the dead which mattered, an interest which extended to funeral procedures as well as to the avenging of homicide."[28]

As to (4) in particular, the belief that the dead must receive proper burial was a long-standing one; it was a divine decree that King Creon in *Antigone* ignored only at his own peril. (For the Greeks as for the Romans, the preferred burial methods alternated between cremation or inhumation.) Homer, for example, has the ghost of the unburied Elpenor implore Odysseus to see to it that his necessary burial rites be performed, "lest I become a cause of the gods' wrath against you."[29] Also, by the fifth century, the Athenian law of homicide evidences this belief in the possible ill will of the dead who are not given prompt burial. Finally, the remaining religious motivations for the urgency of ceremonial burial were chiefly these: (a) that the dead had to be cut off entirely from the sight of the gods in order to avoid polluting or offending the gods and their altars; and (b) that the dead were in need of burial as a last gesture of kindness from the living, without which the dead were not assured admission to the lower world.[30] These, then, were among the principal religious considerations, apart from considerations of public sanitation and health, that lent a strong sense of propriety and proper care on the part of the living to both the corpse and the cherished memory of their dead.

So far I have identified three Greek perspectives on the state of being dead. These are: (1) chthonic personal immortality; (2) divine personal immortality; and (3) natural personal dissolution. In addition, some of the more common attitudes of the living toward the dead have been explored for their religious and moral implications. Attitudes toward the dying process that have come into view include the heroic attitude, plus the idea of euthanasia (about which much more will be said in Chapter Seven). I shall now turn to a brief examination of some of the thoughts of Plato and Aristotle on the prospect of attaining human immortality. This will be followed by a look at what some have called the death-inspired Hellenistic philosophy of Epicureanism.

PLATO AND ARISTOTLE

If I were to summarize in a few paragraphs what Plato and Aristotle had to say about the prospect of attaining human immortality, the story I would tell is this. In the case of Plato, there is every reason to believe that he regarded such a prospect as highly probable. I say "probable" because, despite the elaborate proofs for human immortality offered in his *Phaedo* and elsewhere, we are typically reminded (as at *Phaedo* 107b 5) that the original assumptions on which these proofs depend "still need more accurate consideration." I believe that this sort of qualification signifies just here not only the Socratic habit of leaving inquiries open-ended. It also signifies the doubts still lingering in Plato's mind regarding the possibility of being able to prove *with certainty* the case for divine personal immortality. (That Plato wanted to believe in this sort of personal immortality and most assuredly did, I do not dispute.)[31] If so, then what seems fair to say is that Plato regarded the divinity and personal survival of the soul (which he equated with the essential, true self) as highly plausible but not beyond all doubt.

The significance of his proofs in the *Phaedo* and elsewhere may in the final analysis lie in the fact that they served to demonstrate to many of the skeptical intellectuals of Plato's day that it was at very least reasonable for an intelligent person to adopt and defend belief in divine personal immortality.

The skeptical view, which often took the form of belief in what I have termed natural personal dissolution, Plato definitely opposed. This last view is characterized by Cebes in the following speech at *Phaedo* 60 a:

... [Y]our statement, Socrates, ... seems excellent to me, but what you said about the soul leaves the average person with grave misgivings that when it is released from the body it may no longer exist anywhere, but may be dispersed and destroyed on the very day that the man himself dies, as soon as it is freed from the body. That as it emerges it may be dissipated like breath or smoke, and vanish away so that nothing is left of it anywhere.[32]

Against this view, Plato asserts that the moment of death occurs when the immortal and imperishable soul is freed from the mortal and perishable body. Following death, Plato suggests that the soul is subject to a temporary cycle of reincarnations. The soul may be expected to acquire various kinds of bodies representing lives of lower and higher moral worth. In the *Republic*, it is suggested further that, since we dwell in an essentially just universe, those souls which attain the highest moral virtue (predicated ultimately on acquaintance with Plato's so-called Forms) may one day reside in pure and disembodied splendor among the heavenly stars.[33] Finally, for Plato, as for Pythangoras before him, doing philosophy was regarded as the best way to purify and temper the soul in preparation for that endless odyssey on which it would embark at death.[34]

Aristotle's basic thoughts on the question of human immortality are not quite so easy to interpret as Plato's.

There is at least one passage in his writings, namely *De Anima* 430a 14–25, where he seems to allow the possibility that man's reason (*nous*) might in some sense survive at death. Elsewhere in that same treatise, however, he defends in unmistakable language three propositions that jointly refute the claim that Aristotle believed in personal immortality in any relevant sense. These three propositions are: (1) that the soul is inseparable from the body and is as permanently bound to it as the power of sight is to the eye;[35] (2) that the doctrines of transmigration and reincarnation, similar to those suggested by the Pythagoreans and Plato, are pure myths;[36] and (3) that man, like most other living beings, survives past death only in the narrow biological sense that he may be thought of as continuing his existence in his offspring. At *De Anima* 415a 27, Aristotle puts this last point, which I regard as his basic position, as follows:

... for any living thing that has reached its normal development and which is unmutilated, and whose mode of generation is not spontaneous, the most natural act is the production of another like itself, an animal producing an animal, a plant a plant, in order that, as far as its nature allows, it may partake in the eternal and divine. That is the goal toward which all things strive, for the sake of which they do whatsoever their nature renders possible ... *Since then no living thing* is able to partake in what is eternal and divine by uninterrupted continuance (for nothing perishable can ever remain one and the same), it tries to achieve that end in the only way possible to it, and success is possible in varying degrees; so it remains not indeed as the selfsame individual but continues its existence in something *like* itself, not numerically, but specifically one.[37]

I shall call Aristotle's perspective *species survival*. This raises the question: how did Aristotle look at having to die and the dying process itself? One finds that he supplied opinions which are remarkable for their candor and psychological insight. On conquering the fear of death, he writes:

Fear may be defined as a pain or disturbance due to a mental picture of some destructive or painful evil in the future. Of destructive or painful evils ... only such as amount to great pains or losses. And even these, only if they appear not remote but so near as to be imminent: we do not fear things that are very way off: for instance, we all know we shall die, but we are not troubled thereby because death is not close at hand.[38]

Elsewhere, in the *Nicomachean Ethics*, he admits that, while having to die may be regarded as a profound evil, it is one that can be met nobly and with dignity by the brave:

Now *death is the most terrible of all things*; for it is the end, and nothing is thought to be any longer either good or bad for the dead. But the brave man would not seem to be concerned even with death in *all* circumstances, e.g., at sea or in disease. In what circumstances, then? Surely in the noblest. Now such things are those in battle; for these take place in the greatest and noblest danger .... Properly, then, he will be called brave who is fearless in face of a noble death, and of all emergencies that involve death.[39]

These passages bear the marks of one who gave keen thought to what we would today call the psychology of death and dying. Nor does it escape our attention that, despite the fact that the only sort of survival Aristotle unambiguously identifies is that impersonal sort carried on by one's species, he appears to find some comfort against death by adopting the moral virtue of courage. Perhaps this is the ultimate psychological remedy against the fear of dying. But, in addition, Aristotle declares that "death is the most terrible of all things." Yet, since he also affirms that "God and nature do nothing without a purpose," it is reasonable to conclude that the fact of death did not prevent him from crediting human existence with important meaning.[40] In a word, for Aristotle as for Plato, death did not succeed in rendering life absurd as some existentialist philosophers of this past century have harped. This despite the fact that, as I have argued, Aristotle dispensed with Plato's hope for divine personal immortality.

Yet one may wonder just how satisfying the average Greek or Roman found the differing answers of either Plato or Aristotle on the question of survival after death. I have suggested that, if Aristotle can be said to have provided a psychological remedy to the fear of dying, it lies in the acquisition of a brave character, aided by the intellectual consolation that life and death have some definite purpose. But to the average Greek, trying to become suitably brave in middle life, say, may have seemed a rather empty piece of advice. So, too, Plato's doctrine of divine personal immortality, not to mention Homer's chthonic immortality, may have triggered in the minds of many another unsettling worry: If I survive bodily death, what torments await my soul in the afterworld?

EPICUREAN INSIGHTS

Partly in response to these unallayed fears associated with concern over existence in some afterworld, the materialist philosopher Epicurus (341 – 270 B.C.), influenced by the fifth century doctrines of the atomists Leucippus and Democritus, attempted to refine and popularize the logical and psychological implications of a thoroughgoing atomic theory. Like Democritus before him, Epicurus depicted being dead as an extreme form of natural personal dissolution, namely oblivion. The great strength of Epicurus' position on death was that his theory – according to which death marked the final limit of a human's experience – *prima facie* seemed to fit the universally recorded and manifest fact that "All men are mortal."

Moreover, whereas Democritus, Plato, and Aristotle had allowed that knowledge could be an end in itself, Epicurus adopted a more pragmatic and therapeutic stance. He declared that knowledge was no more than a "remedy for the soul." In fact, according to Epicurus the soul is one and the same as the body. Given this, the entire person is ultimately comprised of tiny, invisible, indestructible pieces of matter called atoms. Thus all functions of the human organism are explained as resulting from atomic processes, and matter is credited with somehow containing the seeds of life.[41]

It comes as little surprise that given this materialistic account of man and nature, Epicurus defines death in purely physical terms as the dissolution and dispersion of the soul atoms into the infinite void. This happens at the exact moment at which all personal sensation permanently ceases. Following Democritus, Epicurus believed that atoms and the void are all that fundamentally exist. (Even the gods, according to Epicurus, were comprised of atoms, though they were not the usual gods of Homer and were said to be totally detached from human affairs; at best the gods were useful only as exemplars of the complete peace of mind (*ataraxia*) at which humans should aim.)[42]

At the heart of his philosophy, which his contemporaries often criticized as a brand of atheism by default, Epicurus offered several fundamental doctrines. These included the so-called Four Remedies of the soul, which declared: "[1] We need not fear God; [2] Death means absence of sensation; [3] The Good is easy to obtain; and [4] Evil is easy to bear." To these four canonical beliefs, he adds the caveat: "vain is the word of a philosopher which does not heal any suffering of man."[43]

Futhermore, the good for man is said to be pleasure, on Epicurus' view, and the highest good a life of secure and lasting pleasure. This lasting pleasure can be gotten through the cultivation of intelligent choice predicted on practical wisdom (*phronesis*). "Practical wisdom measures pleasures against pains, accepting pains that lead to greater pleasures and rejecting pleasures that lead to greater pains. It counts the traditional virtues (justice, temperance, courage, etc.) among the means for attaining the pleasant; they have no other justification."[44]

If there can be said to be a recurring impediment to the peace of mind (*ataraxia*) so valued by Epicurus in his indeterministic, mechanistic universe, it is the fear of death. This fear can tarnish whatever enjoyments one might otherwise achieve in one's life. Hence, in view of this diagnosis, Epicurus' prescription is that such fears can be intelligently removed since they are at bottom totally unfounded. In his letter to Menoecus, among his most important and characteristic thoughts on what death is and how one may end fears of dying, are powerfully summarized:

Become accustomed to the belief that death is nothing to us. For all good and evil consists in sensation, but death is deprivation of sensation. And therefore a right understanding that death is nothing to us makes the mortality of life enjoyable, not because it adds to it an infinite span of time, but because it takes away the craving for immortality. For there is nothing terrible in not living, so that the man speaks idly who says that he fears death, not because it will be painful when it comes, but because it is painful in anticipation. For that which gives no trouble when it comes is but an empty pain in anticipation. So death, the most terrifying of all ills, is nothing to us, since so long as we exist death is not with us, but when death comes, then we do not exist. It does not then concern either the living, or the dead, since the former it is not, and the latter are no more.[45]

Close inspection of the above passage reveals that the force of Epicurus's argument turns on the claim that since being dead and having sensations of any sort are mutually exclusive, and since all good and evil involve sensation, being dead is nothing to fret about. In other words, it is not intelligent to spoil one's life by harboring fears which are in reality baseless.

As seductive as this argument appears to be at first glance, however, I must register at least one objection. It seems to me a doubtful statement that " . . . there is nothing terrible in not living . . . . " This statement may apply for those who were not brought into existence in the first place. But the relevant case is those persons who are now fully alive and in the middle of various projects and human relationships to which they accord value.[46] Death cuts short these ongoing projects and friendships of the living. Thus, the disvalue death brings does not seem entirely owned up to by Epicurus in the preceding passage. At best, he succeeds in thwarting the fears of those who worry about what will happen to them in the afterworld. But to those who highly value their ongoing projects and their friendships with the living, and so bemoan precisely having to give up what they have got and can never reclaim, his argument seems unconvincing and oversimplified.

There were nonetheless many adherents who found consolation in the words of Epicurus and were attracted to his emphasis on philosophy as the medicine of the soul. Like the competing materialistic philosophy of Stoicism which also sought to reduce the fear of death,[47] Epicureanism enjoyed wide popularity in Hellenistic and later Roman times. Its success owed not only to the renewed attention it received from the Roman poet Lucretius, but also to its promise to restore to the individual a level of personal stability and tranquility increasingly valued in the turbulent and expansive Post-Alexandrian age.

SUMMARY

We have now come to the end of this investigation of Greek attitudes toward death. My research points to a plurality of perspectives rather than to any single one. As for how the Greeks pictured being dead, four distinct perspectives have been identified. There are: (1) *chthonic personal immortality* associated with the Homeric religion; (2) *divine personal immortality* associated with the philosophies of Pythagoras and Plato; (3) *natural personal dissolution* associated with Thales, Anaximander, and Anaximenes, and, as we have just seen, Epicurus; and what may be called (4) *species survival* associated with Aristotle.

As for Greek attitudes toward the dying process, at least four perspectives have been discovered. These are (a) *the heroic attitude* according to which one's willingness to die is seen as the best test of one's commitment to the highest sorts of ideals; (b) *the merciful attitude* according to which a quick and easy death (euthanasia) is seen as the best solution to relentless personal suffering; (c) *the Aristotelian attitude* according to which the personal cultivation of courage is seen as the best psychological and moral remedy for meeting death; and (d) *the Epicurean attitude* according to which fears of death can be eliminated by recognizing the baseless and illusory character of these fears.

In addition, the moment of death was shown to receive diverse characterizations by the Greeks. In the Homeric epics, death tended to be characterized as occurring at the moment when the ghost or breath escapes from the mouth or wound of the corpse. For both Pythagoras and Plato, it occurred at the moment that the soul is released from the body. For Anaximenes, it occurred when breathing ceased. In addition, for Aristotle it occurred when the vital life functions (especially heartbeat) ceased. Lastly, Epicurus specifies the moment of death as the permanent loss of sensation.

Amidst this variety of competing beliefs concerning being dead, dying, and the moment of death, can the typical fifth century Hellenic physician be legitimately characterized as favoring some of these beliefs over others?

With one exception, the answer appears to be no. Like other members of Greek society, physicians were free to adopt their own attitudes and beliefs on these matters. It is reasonable to infer, therefore, that almost all of the preceding perspectives were represented among the physicians. Indeed, the Hippocratic treatises, including the Oath, attest to the fact that some physicians were religious or at least respectful of the gods. No doubt a fair number held out hope for the afterworld, though few were in the habit of trying to impose their personal views on their patients. This, even though they regularly divorced their theories of health and medical treatments from supernatural explanations. The lone exception to their diverse personal views was that the moment of death was generally understood to occur when breathing, heartbeat, and sensation collectively and permanently ceased.

Thus far at least three important points have been established in Part One of this study. First, the typical Greek physician of the fifth century was an unlicensed craftsman holding a rather low social rank; he occasionally encroached on the prerogatives of philosophers by telling citizens how they ought to live. I have argued that this encroachment was resisted by most philosophers. Philosophers handily exploited the general knowledge of citizens on health matters by developing the powerful analogy that philosophers were physicians of the soul and so stood in higher authority.

Second, unlike the Egyptians, the Greeks succeeded in freeing medical theories of health and disease from magic and superstition. The impetus for this came largely from the Pre-Socratic philosopher-scientists who sought to put all phenomena on a naturalistic footing and account for all processes in terms of impersonal, universal laws. In addition, the seven principles of the humoral theory of disease were shown to have influenced selected parts of Plato's, Aristotle's, and Galen's theories of health and disease. It was further argued that Galen's eclectic theory of constitutional types integrated some facets of an ontological disease model with those of a physiological disease model that remained his favorite, and which was associated with the humoral theory of health.

Finally, it has been shown that there is no warrant for the mistaken view that the Greeks possessed anything like a single, basic outlook on death and dying. Let us then turn at once to the related consideration of how Greek medical ethics came into being. For despite the fact that the Greeks themselves held a multiplicity of opinions about the nature of death and the value of life, it is puzzling to observe that the ethical values expressed in the Hippocratic Oath demonstrate a singularly high and uncompromising regard for the preservation of human life. To make sense out of this unique feature of the Oath is the task that awaits us in Part Two.

PART TWO

THE RISE OF MEDICAL ETHICS

CHAPTER FOUR

WHO WAS HIPPOCRATES?

Of all the medical writings associated with the name of Hippocrates, probably none is better known to the modern reader than the Hippocratic Oath. Since it is the explicit aim in this investigation to identify and explore ancient philosophical perspectives on abortion and euthanasia, and since the Hippocratic Oath appears to absolutely prohibit these measures, the Oath stands as a natural starting point for the present section.

A number of important questions can be raised. What, if anything, is truly unique about the Oath? How was it different from the literature produced by neighboring non-Greek cultures? Who wrote the Oath, and what purpose and influence did it exercise among physicians and laymen in Pre-Christian Greco-Roman times?

In addition, it will be necessary to examine the relation of the Oath to the Hippocratic Collection as a whole. One wants to learn whether the moral imperatives expressed by the Oath are in basic accord with the medical practices and constraints expressed elsewhere in that sizeable and diverse body of writings.

To round out this investigation into the emergence of medical ethics, attention will also be given to those lesser known Hippocratic treatises which deal primarily with medical etiquette. I shall argue that there is a distinct, logical compatibility between the Oath and these latter treatises: taken together, they display an unmistakable and sustained concern for the patient's best welfare.

MEDICAL ETHICS BEFORE HIPPOCRATES

Darrel Amundsen has written: 'Medical ethics ... is even less apt to be borrowed by members of one society, from another culturally alien to it, than are its medical theory and concomitant technique.'[1] In view of the ethnocentric tendencies characteristically manifested in the moral theories of almost all cultures, no doubt Amundsen is right. But it would be an unfortunate mistake to completely ignore on this score such evidence of medical ethical values that scholars and others have culled from the Near East and Egypt. Assyria, Babylonia, Persia — these were neighboring Eastern cultures with medical traditions well antedating mainland Greece itself. And to the South, few can any longer seriously doubt the cross fertilization that occurred as a result of trade between the Greek city-states and Egypt.

To overlook that the Greeks may have acquired useful ideas from their widespread commercial encounters along the Mediterranean sea channels or on overland expeditions — ideas possibly affecting their art and medicine, if not their philosophy — is to indulge an unaffordable provincialism.[2] In fact there is little to risk in looking at the larger ancient landscape and much to gain. For how else can one accurately gauge the originality of the Greek contribution to Western civilization, and in particular ethics, if one hesitates to survey her ancient neighbors? Essentially, then, I agree with Amundsen: the probabilities of inter-cultural influence in the field of medical ethics are slim. But I would insist that the moral traditions of neighboring cultures hold more than "peripheral relevance" to the fullest understanding of the unique character of Greek medical thinking.[3]

For example, unlike fifth century Athens, in Babylonia and Egypt medicine was practiced subject to strict state regulation. Attention has already been given to the testimony of Aristotle according to which an Egyptian physician is warned by the state not to alter a course of treatment until at least four days have passed, or else be subject to legal penalties.[4] But the legislative regulation of medicine appears to have advanced even further in ancient Babylonia. The Code of Hammurabi (dating from about 1727 B.C.) contains the first recorded attempt by any culture to protect patients from incompetent doctors.

The medical laws of Hammurabi may be divided into two broad categories. Some refer to the medical care of noblemen. Others pertain to commoners or slaves. In the case of noblemen, the Code stipulates that if a surgeon performs surgery with a bronze lancet or specifically operates with a lancet on the eye, and this results either in the loss of the eye to the patient or the patient's death, then the state shall cut off the surgeon's hand (Code #218).

In contrast, if the physician operates with a bronze lancet on a slave owned by a commoner and the result is that the slave dies, the physician must replace the slave by one of equal value (#219). Moreover, if as a result of a surgical operation on a slave's eye the slave loses his sight, the physician must pay the commoner one-half the market value of his damaged slave (#220). Add to the preceding regulations those provisions of Hammurabi's Code that set differential medical fees for surgery based on the socio-economic status of each patient (#215–217; 221–223), and the picture emerges of a surprisingly comprehensive system of legislative sanctions defining the proper conduct of surgical medicine.[5] It is important to note, however, that Hammurabi's Code remains silent on most non-surgical medical procedures. Hence no penalties are specified for dietary and other forms of therapy. This was probably due in large part to the fact that surgery was perceived to carry more immediate and permanent risks to the patient compared to other forms of medicine known to the Babylonians (including magic). Therefore legal sanctions against the possible abuses of surgical procedures were deemed especially necessary.

There is the related question of whether these diverse Near Eastern cultures defined the favored characteristics of their ideal physician in their medical or religious literature. All that can be said is that in the case of the Assyro-Babylonian and Egyptian cultures no such literary evidence has survived. Obviously this does not rule out that they possessed such idealized medical standards.

Investigations into the medical ethical values of ancient Persian culture have proved more fruitful. Amundsen calls attention to a sixth century A.D. passage taken from the third book of the Sassanian Persian's encyclopedic *Dinkard*. This passage characterizes in very specific terms the qualities of the ideal physician. The importance of this passage is that some scholars think it may be based directly on a lost portion of the much older *Zend-Avesta*, of which only about one-third remains.[6] For comparative purposes in relation to the Greek ideal, it is interesting to read that according to the *Dinkard* the best physician

... should know the limbs of the body, their articulations; remedies for disease; ... should be amiable without jealousy, gentle in word, free from haughtiness; an enemy in disease, but the friend of the sick, respecting modesty, free from crime, from injury, from violence; expeditious; ... noble in action; protecting good reputatuon; not acting for gain, but for a spiritual reward; ready to listen; ... possessed of authority and philanthropy; skilled to prepare health-giving plants medically, in order to deliver the body from disease, to expel corruption and impurity; to further peace and multiply the delights of life.[7]

Though one can be struck by the modern ring of the preceding description, it must be kept in mind that the "physician" to whom these words refer was a practitioner of magic in the Persian culture. That is, Persian physicians were also Zoroastrian priests. Thus, the Persians, like their Egyptian and Assyro-Babylonian counterparts, freely mixed the rational elements of medical practice with supernatural magico-religious elements. Yet it is exactly this tendency that the Hippocratic physicians opposed. For the Greeks succeeded in divorcing priestly functions from their own concept of the ideal physician.[8]

On the issue of euthanasia, there exists no direct evidence which unambiguously reveals the moral stand taken in the Near East. It *is* known that in both Egypt and Assyro-Babylonia, a person who committed *suicide* was thought to have irrevocably cut himself off from the gods.[9] In view of this, it seems plausible to infer that on religious grounds, physicians from these cultures may have had some serious reservations about encouraging or contributing to active voluntary euthanasia. But this is conjectural, and presupposes more than can now be proved. That is, it presupposes that the act of suicide would have been understood by these ancient peoples as morally equivalent to voluntary enthanasia. Personally, I have my doubts.

Nonetheless, it is true that in the *Edwin Smith Papyrus*, as well as in the *Papyrus Ebers*, the Egyptian physician divides his cases into three groups. If the prognosis he expected was favorable, he commented "an ailment I shall treat." On uncertain prognoses, the comment is "an ailment I shall combat." But if the prognosis is unfavorable, the notation reads "an ailment not to be treated." Yet this last category of patients was not abandoned. As some of the cases explicitly reveal, the physicians of Egypt took steps to at least relieve the suffering; they helped the patient remain as comfortable as possible till death.[10]

As for the Persians, still less can be established. If the above passage from the *Dinkard* is carefully examined, there appears nothing in it that explicitly excludes mercy killing. Those portions of it that describe the ideal physician as the friend of the sick, free from crime, the bringer of peace, the multiplier of the delights of life, and so on, are not necessarily incompatible with a pro-euthanasia perspective. This is especially so if voluntary euthanasia is viewed as a kindness to the patient. In other words, as a benevolent, not a criminal act; an act that brings peace to a life judged no longer capable of delight.

In view of this possibility, I cannot agree with Amundsen's conjecture. Amundsen argues that the ancient Persian doctor, just because he was also a priest and interpreted disease as caused by the forces of evil, would probably *not* have hastened the end of hopeless patients.[11] This may be doubted. For *if* the patients were indeed perceived by the priest as hopeless, and if they begged for death and an end to their suffering, who can deny that the priest-physician might have been motivated by religious compassion and mercy to assist the patient in ending his torment? I do not want to press this objection, but it does serve to illustrate the ambiguity of the Persian position on this issue.

Regarding the practice of abortion, somewhat more is known. In Persia, the sacred book known as the *Vendidad* prohibited abortion. "The woman, the nurse who performed the abortion, the woman's father or the father of the fetus, if the abortion was sought at their instigation, were all held guilty of willful murder."[12]

Similarly, in Assyria there existed laws against abortion possibly as early as the fifteenth centruy B.C. One Assyrian law from that period provided that "if a woman had an abortion by her own act, whether or not she survived the ordeal, she was to be impaled on a stake and left unburied."[13] However, no mention is made in Assyrian law of penalties against physicians or midwives who may have aided women in their abortion attempts. This is indeed puzzling. It is possible that those who conspired with the women were implicitly covered by this edict.

Lastly, in the case of Egypt the medical papyri *do* contain prescriptions concerning how to induce abortions. This implies that abortions were almost certainly performed by Egyptian doctors. But whether it was entirely legal or considered moral, and to what extent and under what conditions it was understood to be medically warranted, must at present remain a matter of speculation.

In sum, it seems fair to conclude that the latest evidence permits only the sketchiest picture to be rendered regarding the medical ethics of the ancient Near East and Egypt. It is worth emphasizing that, unlike the Greeks, so far no literature appears to have survived from the four ancient cultures under discussion which is specifically devoted to medical ethics. If, in fact, any Assyrian, Babylonian, Persian, or Egyptian doctors or priests wrote an ethical code relating to medicine, we have no evidence to this effect. We possess only the most tentative idea of what values they might have endorsed. Even so, the mosaic of laws and religious prescriptions at our disposal reveals an overall and varied interest in the basic welfare of the patient, whether he be freeborn or slave. It is evident that the physician-priests from these neighboring cultures had a definite sense of the right and wrong way to care for those seeking their services. This sense derived mainly from the authority of their laws or their religion. But they were by no means free agents practicing their craft independently of state interference, as were their Greek counterparts.

THE PUZZLE OF HIPPOCRATES' IDENTITY

Before turning to the Hippocratic Oath, it is appropriate to comment on the puzzling fact that very little is known about Hippocrates himself. The only extant ancient sources that mention him by name and which were composed near the period of his life (469–399 B.C.), are two passages in Plato. One is from Plato's *Protagoras*, the other from the *Phaedrus*. In the generation following Hippocrates' death, there exists a passing remark by Aristotle in his *Politics*, and the medical writings of Meno. Lamentably, we have little to go on (even though he was the exact contemporary of Socrates).

In the *Protagoras* passage, it is attested that (a) Hippocrates was a native of the Greek island of Cos; (b) he was apparently as well known in his own day as the celebrated sculptors Phidias and Polyclitus; and (c) he accepted payment for instruction in the art of medicine.[14] In addition, the *Phaedrus* establishes that Hippocrates practiced medicine based on an organistic principle.[15] To him is attributed the principle that the nature of any component part or organ can only be properly understood after thoroughly examining the whole to which it belongs. In the *Phaedrus* passage, the discussion turns on a Socratic analogy between rhetoric and medicine. Hippocrates' organistic approach to medicine is cited approvingly by Socrates. By reference to it he reinforces his point that both the master rhetorician and the master physician must study the whole nature of the special objects of their arts (these objects being the body and soul, respectively). Such a fundamentally comprehensive approach to inquiry was much favored by Socrates, who, in the *Phaedrus*, endorses it for anyone who " . . . mean[s] to be scientific and not content with mere empirical routine . . . ."[16]

Moreover, Aristotle mentions Hippocrates by name just once.[17] In his *Politics* he identifies Hippocrates as a great physician and implies that he was short in stature. His point is that Hippocrates' greatness stemmed from his excellence as a physician who performed well *qua* physician; Hippocrates' greatness was not merely due to noble birth or wealth or physical size. It derived rather from the quality of his work above all else.

Furthermore, Aristotle's suggestion that Hippocrates was a man of short stature is the only reliable physicial description that we have. While efforts have been made over the centuries to identify various ancient statues (mostly copies of Greek originals) with Hippocrates of Cos, no positive identification of Hippocrates' likeness has so far proved convincing. Albert Lyons has argued that some first century A.D. Roman coins that have recently been unearthed from the island of Cos and which bear the name of Hippocrates reveal " . . . the likely appearance of Hippocrates . . . ."[18] This is extremely conjectural, however, in view of the absence of other ancient artifacts against which this alleged image may be checked. Nor will it serve in this case to argue that since numismatic specimens often obscure the true facial features of their subjects, Lyons' judgment is all the more dubious. For, in fact, before the fourth century A.D., numismatic portraiture was uncommonly realistic. So the decisive question here is, Did the first century A.D. artists themselves any longer know what Hippocrates may have looked like? This question becomes especially pertinent in the case of coins struck 500 years after their subject's death and discovered near a site also popular to ancient collectors and travelers. So while I remain skeptical, Lyons' conjecture cannot be dismissed.

Aristotle's student, Meno, who wrote a doxographical compendium of early Greek medicine, states that Hippocrates held a theory of disease according to which badly digested or unwisely selected food produces putrid air inside the body that invades the organs and tissues, causing disease. This theory of putrefaction is attributed to Hippocrates in the so-called *Anonymous Londinensis,* a work discovered among papyri long stored in the British Museum, and one that contains some of Meno's medical writings.[19] To these few details one can add the likely facts that Hippocrates traveled widely, at least in his native Greece, as did most physicians of his day. Also, that he died in Thessaly and was buried near Larissa around 399 B.C. is most probable. More about his life and personality cannot now be ascertained.[20]

In view of how little is actually known about the historical Hippocrates, how can one square this modest account with the later, unbounded fame Hippocrates acquired in antiquity as the Father of Medicine? This is a puzzling question and no simple story can be told.

In my opinion, the most credible answer appears to be this.[21] To his Greek contemporaries Hippocrates *was* regarded as a famous physician, but he was one among many. For example, in his medical history Meno discusses twenty famous Greek doctors. Hippocrates is *not* singled out as more important than the rest. Also, Plato refers to a number of other doctors in his dialogues and he seems to imply that they are also approximately as famous and well known to his readers as Hippocrates.

In addition, it is likely that scholars at the great library at Alexandria, in Egypt, began assembling and seeking out medical texts associated with the teaching of Hippocrates at Cos, along with materials from other famous Greek medical centers, sometime during the fourth centuty B.C. They probably began with a fair number of *anonymously authored* fifth and fourth century Greek manuscripts. When their interest in identifying the works of Hippocrates peaked on account of the latter's growing reputation among his many medical disciples, the Alexandrian scholars or their assistants probably began crediting the most impressive of these anonymous manuscripts to the physician from Cos. If the core of the Hippocratic Collection did in fact

form in this or some related way, perhaps it continued to enlarge owing to
these occasional irregularities in manuscript classification. Thus, it is not hard
to imagine that in succeeding centuries the *name* of Hippocrates exceeded
the fame and stature of the already prominent life that the real Hippocrates
had in fact lived.

What is more, to this account three fortuitous developments bearing on
the fame of Hippocrates must be included.

First, the Romans of the Imperial period enshrined fifth century Greece
as the golden age. Greek writers and thinkers of that idealized period were
regarded by most Roman intellectuals as matchless. For example, the Roman
Erotian (fl. 50 A.D.) ranked Hippocrates the equal of Homer as a writer.[22]
This sort of romanticized Roman estimate of the Greek golden age doubtless
redoubled Hippocrates' fame. Hippocrates was, after all, by Erotian's day
the best known of all the fifth century Greek physicians in Nero's Rome.
Again, this owed largely to the labors of the Alexandrian scholars who aimed
at amassing the entire sum of human knowledge in a single library despite the
attendant risks of misattribution.

Second, while Galen himself questioned the authenticity of some of the
Hippocratic manuscripts, he personally revered Hippocrates as the ideal
physician. In this, he urged fellow physicians to follow his lead.[23] This was
indeed fortunate. It may be said that almost any craft or profession seeks
to identify an ideal hero who is the embodiment and reminder of the values
and aims it holds dear. (Socrates has often been cast in such an idealized role
by professional philosophers, for example.) But the personal endorsement of
Galen could not have brought to the name of Hippocrates a more enduring
and powerful certification of his worth for centuries to come.

Third, it cannot be overlooked that the ethical values expressed in the
Hippocratic Oath conformed in many ways to the respect for human life
ethic which was championed by the latest of the ancient religions, namely,
Christianity. To this last point I shall return in the subsequent chapter. But
by way of anticipation, it is evident that the acceptability of much of the
Oath popularly credited to Hippocrates by the early Church Fathers themselves, did much to sustain his legend.

Before turning directly to explore the Oath, I must stop for a moment
to consider the problem of the many stylistic and logical inconsistencies
contained in the Hippocratic Collection as a whole. Because of the scarcity
of evidence which has so far rendered inconclusive efforts to pinpoint
any single one of the Collection's more than 60 books as a genuine work
by Hippocrates, a rather recent and sensible convention has arisen among

contemporary writers that I shall continue to employ in the remainder of this study. This is the convention of referring to the entire collection as the product of frankly anonymous "Hippocratic authors." In so doing, however, I do not wish to exclude the possibility that Hippocrates may have actually written at least some of these works. On that question I urge an open mind.[24]

CHAPTER FIVE

THE HIPPOCRATIC OATH

At first sight it is indeed ironic that the so-called Hippocratic Oath, which is the most renowned medical ethical document and the one most popularly associated with Hippocrates' name, is now judged by very few scholars to be authored by Hippocrates. What's more, it is especially doubtful that the Oath accurately reflects the ethical values and medical practices which the Hippocratic authors favored and typically followed in their practice of medicine. In what follows, I shall undertake to argue for these two basic conclusions. My plan will be met in two steps. First, I shall critically discuss two important contemporary positions on the date, origin, and purpose of the Oath. Then I shall argue that the Oath represents essentially an esoteric ethical code which is partly, though not exclusively, of Pythagorean origin.

The Oath itself is a very short document. It reads in full:

Oath

P1   I swear by Apollo Physician and Asclepius and Hygieia and Panaceia and all the gods and goddesses, making them my witnesses, that I will fulfill according to my ability and judgment this oath and this convenant:

P2   To hold him who has taught me this art as equal to my parents and to live my life in partnership with him, and if he is in need of money to give him a share of mine, and to regard his offspring as equal to my brothers in male lineage and to teach them this art – if they desire to learn it – without fee and covenant; to give a share of precepts and oral instruction and all the other learning to my sons and to the sons of him who has instructed me and to pupils who have signed the covenant and have taken an oath according to the medical law, but to no one else.

P3   I will apply dietetic measures for the benefit of the sick according to my ability and judgment; I will keep them from harm and injustice.

P4   I will neither give a deadly drug to anybody if asked for it, nor will I make a suggestion to this effect. Similarly I will not give to a woman an abortive remedy. In purity and holiness I will guard my life and my art.

P5   I will not use the knife, not even on sufferers from stone, but will withdraw in favor of such men as are engaged in this work.

P6   Whatever houses I may visit, I will come for the benefit of the sick, remaining free of all intentional injustice, of all mischief and in particular of sexual relations with both female and male persons, be they free or slaves.

P7   What I may see or hear in the course of the treatment or even outside of the treatment in regard to the life of men, which on no account one must spread abroad, I will keep to myself holding such things shameful to be spoken about.

P8   If I fulfill this oath and do not violate it, may it be granted to me to enjoy life and art, being honored with fame among all men for all time to come; if I transgress it and swear falsely, may the opposite of all this be my lot.[1]

In order to clearly refer to the several provisions of the Oath, I have assigned numbers to its eight paragraphs (hereafter P1–P8). The Oath can be usefully divided into four basic sections. These are: (1) the *preamble* (P1), involving the invocation of the appropriate health-related gods and the actual swearing of the Oath; (2) the *covenant* (P2), involving the duties of the student to his teacher, to his teacher's family, and his obligations regarding the transmission of his medical knowledge to others; (3) the *ethical code* (P3–P7), involving, among other things, prohibitions against dispensing poisons, performing abortions, performing surgery, having sexual relations with patients, and revealing to others details of a patient's private life; and (4) the *peroration* (P8), wherein the speaker expresses his belief that his own good or bad fortunes shall rest in his ability to live up to the Oath's provisions. While all four parts of the Oath have received attention in attempts to fix its date and likely origin, most discussion centers on the meaning of (2) the covenant and (3) the ethical code.

Furthermore, all interpreters of the Oath immediately face a common obstacle: some of the ethical rules embodied in the Oath are logically inconsistent with passages found in other treatises of the Hippocratic Collection. In addition, several rules are historically inconsistent with the known medical practices of the Classical and later Greco-Roman period. For example, several treatises in the Collection recommend surgery. Moreover, the surgical treatises are generally regarded to be the most medically precise and useful parts of the Collection. As for prohibitions against aiding a patient in suicide or abortion, both were practices basically acceptable in the Greek as well as the Roman era. That is, neither religious or legal sanctions, nor moral disapprobation typically forbade such conduct.[2] Also, it is most unlikely that the Oath became a regular topic of debate among physicians or others in Pre-Christian antiquity. There are no clear cross-references to it in the entire Hippocratic Collection. No known references to details of the Oath occur until the first century A.D. writings of Scribonius Largus.[3]

In view of these puzzling facts, modern interpreters of the Oath have characteristically adopted one of two possible interpretive strategies. On the one hand, some, like Savas Nittis, have sought to explain away various troublesome and otherwise irreconcilable features of the Oath in such a way that the Oath can still be regarded as a work genuinely authored by Hippocrates or his closest associates.

On the other hand, following a quite different strategy are scholars like Edelstein. This second strategy gives up the claim that the Oath is a typical Hippocratic writing or that it expresses typical Greek ethical values. Instead, an effort is made to show that the Oath is atypical of Greek medical literature and practice. Thus it is argued that its origin must lie with an individual author or school whose ethical values sharply differed from those of main-stream Classical society.

Edelstein's is currently the most widely discussed and comprehensive theory belonging to this second group. He argues that the Oath is not a writing of the historical Hippocrates. Rather, it is, in his opinion, plausibly understandable solely as the product of the esoteric ethical teachings of fourth century B.C. Pythagoreanism. Its primary purposes, he suggests, were to (a) mutually bind teacher to pupil: (b) keep the soul of the physician pure in accord with essentially Pythagorean religious values: (c) designate the proper moral duties between physician and patient; and, probably most important, (d) reform the practice of medicine.

To Nittis' and Edelstein's opposing interpretations I shall now turn.

THE AUTHORSHIP QUESTION

Nittis claims that the Oath was written by Hippocrates himself in Athens between March and October of the year 421 B.C.[4] His reasoning seems quite tenuous. A persistent flaw in his analysis is that Nittis inexplicably ignores the fact that the Oath is written in the Ionic dialect. Yet if, as he contends, it was likely written by Hippocrates in Athens in 421 B.C., and was used by an active medical guild founded by Hippocrates in fifth century Athens to establish the duties and privileges which obtained between students and teachers of the medical craft, why was it not written in the *Attic* dialect common to Athens, as one would normally expect? This question has all the more force since Nittis also claims that the covenant (P2) originally had quasi-legal status *in Athens*. But, if so, it would normally have been written in Attic, as were other Athenian legal and quasi-legal documents of this sort. Nittis additionally claims that the covenant was typical of the covenants of most fraternal guilds in Classical society. But he does not present a single scrap of evidence that in such guilds the pupil virtually adopted his teacher as a parent, as the Oath seems to imply.

In addition, on the basis of a very ambiguous piece of historical evidence taken from Thucydides' description of events in the year 421 B.C. (which may have involved no more than an arbitrary choice of words), Nittis infers that the Greek words for covenant (*xyngraphe*) and law (*nomos*) were used interchangeably in Athens between March and October, in 421 B.C.[5] He also alleges that these two words are likewise being used interchangeably in the Oath's second paragraph, thereby linking its authorship to that very year. Yet this alleged equivalence is not at all proved or even plausibly suggested, so far as I can determine, from the relevant passages he points to in Thucydides (see *The Peloponnesian War* VII, 67 and 97).

Yet it must be admitted that a careful reading of these two passages does not specifically exclude Nittis' interpretation. But when one adds to this already questionable claim of word synonymity the additional dubious claim on which the second leg of Nittis' theory stands, namely, the claim that Hippocrates was probably in Athens between March and October in the year 421 B.C., it appears that Nittis' interpretation assumes too much. For Nittis apparently assumes that the brief mention of Hippocrates' name by Plato in *Protagoras* and *Phaedrus* implies that Hippocrates " . . . was living in Athens, enjoying great reputation, not only as a practicing physician, but also as a teacher of medicine . . . " in 421 B.C. But Nittis nowhere argues for the dramatic date or historical context of these two dialogues.[6] If he is aware of the controversies surrounding the exact dating of individual works from the Platonic Dialogues, he does not share his thoughts with his readers as might be hoped.

At bottom, then, Nittis' interpretation on the question of authorship is unpersuasive. This is so primarily because he appears to underrate or ignore the complexity of the very premises on which the bulk of his principal claims depend.

THE PROHIBITION AGAINST SURGERY

In a related article, Nittis tries to resolve the inconsistency between the Oath's prohibition against surgery (P5) and the outright acceptance of surgery in other Hippocratic treatises. This he seeks to do since he identifies Hippocrates as the probable author of the Oath and is at the same time sympathetic to the view that Hippocrates or his disciples authored most other treatises in the Collection. This logical inconsistency involving surgery is thus an embarrassment to the unity of his theory. Nittis advocates the view, first reluctantly suggested by Littré, that the words *ou temeo*, herein translated "I will not use the knife," have a hidden meaning. These words, he argues, " . . . must in fact refer to an abomination, such as castration was considered among the Greeks, since *temnein* means also 'to castrate.' "[7]

Traditionally, scholars from the Renaissance to the nineteenth century had interpreted this troublesome fifth paragraph to signify that the Oath intended to draw a line between surgery and internal medicine. Many interpreters ascribed this separation of tasks to the supposed fact that surgery was in ancient times held to be beneath the dignity of the physician. More recent scholarship, however, has shown this last supposition to be groundless. For example, Littré pointed out that the ancient practitioner was a surgeon *as well as* a physician. Littré considered the earlier mistaken interpretation to be the result of more modern medical prejudices against surgery, and I suspect he is right.[8] Incidentally, in ancient medicine surgery included cautery, the setting of fractures, and cutting; and the Oath explicitly restricts only the last of these.

Another theory was that the operation performed in ancient times for removal of stones from the bladder, the lithotomy, was in many instances fatal. It also may have been in general morally repugnant to the Hippocratics, some argue, since it allegedly rendered men impotent. Are these the likely reasons why the Oath relegates the lithotomy to a distinct group of practitioners specializing in this procedure?

It seems that there are several reasons why such explanations fail. In the first place, all this presupposes that Classical Greek medicine was for the most part divided into specialties. This I have already argued against in Part One. Also, there is no evidence which suggests that lithotomies were usually fatal. In fact, they appear to have been a common operation of modest success. Yet, aside from these considerations, the decisive objection to such explanations is this: If all those who swore by the Oath harbored, for whatever reasons, definite medical or moral reservations about performing lithotomies, would it not have struck them and others as doubly reprehensible to refer their patients thereafter to " . . . such men as are engaged in this work" (P5)? I am inclined to think so. For, as in the cases of abortion and euthanasia (P4), the prohibition against surgery *could* have been made *without* directing patients toward the services of others (say, poison peddlers or perhaps some midwives). This very objection has also given Jones and Edelstein concern for doubt.[9]

To overcome it, Nittis has sought to dissolve this difficulty by translating P5 to read:

I will not cut, indeed not even sufferers from stone, and I will *keep apart* from men engaging in this deed.[10]

As mentioned, he construes the cutting to refer mainly to castration. Castration, he argues, may have been an occasional untoward consequence of lithotomies. Moreover, the second conjunct of P5 is adjusted by Nittis to escape compounding the moral dilemma just referred to of forbidding the Oath-takers to perform lithotomies, yet letting patients undergo them at the hands of others.

However, in order to justify this grammatical maneuver, Nittis must convincingly translate the verb *ekchorreso* to mean "I will keep apart from." Here he faces opposition. The standard senses of this verb do not include the preceding sense that Nittis needs.[11] Edelstein has also objected in this regard that Nittis is unable to adduce *any* evidence for such an unnatural usage.[12] Therefore, despite his otherwise helpful investigation of the history of castration in Greco-Roman times, and despite the admitted possibility that lithotomy did sometimes involve the unintended consequence of castration which the Hippocratics may have wished to avoid, Nittis' translation of P5 is not very plausible. Try as he may to salvage the alleged Hippocratic authorship and logical consistency of the Oath, these and related problems remain.

Edelstein's opposing theory can be conveniently introduced on the heels of this controversy over the prohibition against surgery. The overall form of his argument is that of a lengthy disjunctive syllogism. Arguing deductively, he examines the four main parts of the Oath (see above, pp. 69–70) by repeatedly proposing a string of possible cultural influences which might explain the origin of the Oath's various ethical provisions. In most cases, he *appears* to eliminate all the competing alternative sources of influence save one: the atypical values of the Pythagoreans. I find it admirable that his interpretation, unlike Nittis', is comprehensive; it addresses all the parts of the Oath and avoids thereby the attendant risks of a fragmentary approach. Furthermore, Edelstein's interpretation has gained the endorsement of Sigerist, and, more recently, the qualified assent of Amundsen.[13]

I shall argue, on the other hand, that Edelstein's proposal, while not entirely wrong, is somewhat flawed. I agree that the Oath was probably conditioned by Pythagorean influences. But I by no means share Edelstein's confidence that no other relevant influences informed its contents. Thus it seems that Edelstein has overpressed his evidence, as I shall now show.

Consider, once again, the Oath's prohibition against surgery (P5). Edelstein reminds us in an earlier discussion of abortion and euthanasia that it was considered unethical by the Pythagoreans to knowingly take any course of action that might shed blood or lead to the destruction of human life. (I shall examine the details of their respect for human life ethic in the next two chapters.) Moreover, he then introduces two additional pieces of evidence. The first, from the fourth century B.C. historian Aristoxenus, attests that the Pythagoreans ranked surgery last on their list of preferred therapies. Diet is given first choice; drugs second. "They believed least of all in using the knife and in cauterizing," Aristoxenus is quoted as saying.[14] The second is from Plato's *Timaeus*.[15] Adopting Taylor's view that the *Timaeus'* medical theories are essentially of Pythagorean origin, Edelstein observes that surgery and cauterization are not even mentioned in that dialogue as medical therapies. From this he infers that there must have been Pythagoreans who refused to apply *any surgical treatments whatsoever* — treatments otherwise so widely used in Greek medicine.[16]

Edelstein then concludes that the prohibition against surgery in the Oath really represents a sort of compromise. The Oath *does not allow* surgery, in the sense that the Oath taken is strictly forbidden to perform cutting; yet it leaves open the practice of surgery to " . . . such man as are engaged in this work." Here it ostensibly disagrees with the more rigorous Pythagorean thera-peutics implied by the *Timaeus*. But the Oath *does allow* cauterization, in possible agreement with the *Timaeus* and at least part of Aristoxenus' testi-mony (which, strictly speaking, does not specifically forbid either surgery or cauterization, but only ranks preferences). As to the longstanding and thorny problem which would remain even if all this is temporarily granted, namely, that the Oath-taker apparently does not oppose his patients' visiting the very surgeons whose operations he opposes, Edelstein supplies the following explanation:

. . . the Pythagorean physician will allow others to help his patient in his extremity. The stipulation against operating is valid only for him who has dedicated himself to a holy life. The Pythagoreans recognized that men in general could not observe any elaborate rules of purity; in this they saw no argument against that which they considered right for themselves. To give place to another craftsman, especially in such instances where the patient might fall prey to sinful temptation, i.e., suicide, certainly was a duty demanded by philanthropy, by commiseration with those who suffered.[17]

In my opinion, Edelstein's explanation here is preferable to Nittis'. It does not depend on Nittis' unnatural translation of the text. Also, it addresses more directly the moral implications of the Oath not forbidding patients from seeking out the services of surgeons (a moral option the author of the Oath undeniably had). Yet Edelstein's preceding interpretation of P5 runs afoul of his own strong claim according to which " . . . *only* in connection with Pythagorean medicine . . . " does this prohibition against surgery make sense.[18] This seems to me doubtful.

In the first place, there is at least one other passage in the Hippocratic Collection which ranks surgery and cauterization behind other therapeutic methods (e.g., drugs). The Hippocratic *Aphorisms*, VII, 87, states: "Those diseases that medicines do not cure are cured by the knife. Those that the knife does not cure are cured by fire. Those that fire does not cure must be considered uncurable."[19] In general, surgery was not resorted to as a treatment of first choice by the ancients; the Pythagoreans were therefore *not alone* in ranking it behind other therapies. Hence, even taking Aristoxenus' report at face value, is not the alleged uniqueness of the Pythagorean ranking of surgery at best a matter of degree only? Have we not seen that this kind of trait is expressed by other medical writers as well?

Second, it is fair to question whether the absence of the mention of surgery in the *Timaeus* is equivalent to prohibiting this practice. It could as naturally mean that surgery was not among the *preferred* medical treatments. But, if so, the Oath's absolute stand against its adherents' performing surgery is not fully equivalent to the meaning of the *Timaeus* passage, as Edelstein would have us believe.

Lastly, it is nowhere stated in the testimony Edelstein presents that the Pythagoreans *alone* excluded surgery. Indeed, it is possible that the prohibition against cutting represents a reaction on the part of some non-Pythagorean physicians against a popular fifth and fourth century distrust of surgical procedures.[20] For instance, we recall Heraclitus (c. 544–484 B.C.), who, commenting on the ironic nature of the physician's craft, complained of " . . . physicians, who cut and burn, demand payment of a fee, though undeserving, since they produce the same pains as the disease."[21] By placing a personal constraint on surgery, physicians swearing the Oath might have sought to reduce the public controversy, fear, and misunderstanding generated by the most dramatic and painful therapy at their disposal. But if so, there may be more than exclusively Pythagorean influences at work in shaping the Oath's disclaimer against surgery.

THE COVENANT RECONSIDERED

Returning to the question of the covenant (P2), I find it equally doubtful that Edelstein is entirely right when he claims that " . . . all the other demands enjoined upon the pupil [by the covenant] may likewise be explained only in connection with Pythagorean views . . . " I do not dispute his weaker claim that " . . . at least they are compatible with them."[22] On this point, I agree with C. J. DeVogel. DeVogel concurs with Edelstein's claim that in the Pythagorean school pupils regarded their teachers as adopted fathers, whose welfare they looked after, with whom they shared their possessions – if need be – and performed other special duties as the Oath seems to require. She contends, however, that Edelstein is "definitely incorrect" in claiming that, apart from the Hippocratic Oath, we possess no other non-Pythagorean indications of the existence of such an attitude in Greek culture.[23] I think rightly, she calls attention to the evidence of Plato's *Crito* as a tacit instance of a pupil (Crito) looking out for the personal welfare of his teacher (Socrates). After all, if Socrates and a few other like-minded educators took no money for instruction (unlike the sophists, whom Socrates fiercely opposed on this issue), how, then, did they survive?

Furthermore, DeVogel reminds us that the Epicureans as well followed this pattern of mutual adoption. If so, this fact is all the more relevant. Edelstein himself assigns the Oath to the fourth century B.C., an important century to be sure for the flowering of Epicureanism. Add to these considerations the additional conjecture that popular fifth and fourth century mystery cults also had their novices call "father" those who introduced them to the secret cult doctrine, and one begins to suspect that Edelstein has once again made a hasty conclusion. He presses too narrowly the possible Pythagorean tendencies that appear – alongside *other* possible sources – to account for the special pupil-teacher duties embodied in the covenant.

THE ETHICAL CODE

Having indicated some weaknesses in Edelstein's account of the prohibition against surgery (P5) as well as in his analysis of the covenant (P2), I would like to focus attention on the Oath's ethical code *per se* (P3–P7). Strictly speaking, the restriction against surgery belongs to this code, but it was convenient to discuss it separately given its special place in most disputes about the Oath.

One may divide the ethical code into two parts. The first part concerns rules for the healing of diseases ((i)–(iv), below). The second concerns rules regulating the physician's behavior in all matters not directly connected with treatment, such as the physician's relation to the patient, the patient's family, and so on ((v)–(vi), below). Together, then, the ethical code embodies *six* related imperatives which the Oath-taker promises to follow:

(i)   The imperative to apply dietetic measures in accord with one's best judgment and to practice medicine so as to protect the patient from harm and injustice (P3).

(ii)   The imperative neither to give nor suggest to a patient the use of a deadly drug (P4).

(iii)   The imperative not to give to a woman an abortive remedy (P4).

(iv)   The imperative not to perform surgery (P5).

(v)   The imperative not to have sexual intercourse with the patient or any member of the patient's household, be they freeborn or slave, female or male (P6).

(vi) The imperative to keep secret and not divulge to others the private details of a patient's life which may be acquired during the course, or even outside the course, of treatment (P7).

So far, I have already addressed (iv). I shall now examine Edelstein's interpretation of (i), (v), and (vi), arguing that for the most part in these instances too he exaggerates the Pythagorean influence. The twin imperatives against assisting patients in abortion and euthanasia ((ii)–(iii)) I shall leave for last; they are of crowning importance and introduce the principal topics of Part Three of this study.

It is Edelstein's contention that the imperative concerning (i) dietetics agrees with the dietetic practices of the Pythagoreans and " . . . acquire [s] meaning *only if* seen in the light of Pythagorean teaching."[24] Now it is well enough established that the Pythagoreans depended on elaborate dietary rules, both for the preservation of health and the treatment of disease. There is Aristoxenus' testimony, cited above. There is also the testimony of Iamblichus (250–325 A.D.), according to which the Pythagoreans " . . . approved of dietetics most of all. They practiced it with great precision."[25]

But the second half of Edelstein's contention is mistaken. For example, dietetics was in general a widely used form of therapy by the Hippocratic physicians.[26] It was not exclusively emphasized by the Pythagoreans (as Edelstein himself seems aware). On what additional grounds, then, does he see in (i) the unique impress of Pythagoreanism? His principal justification turns on the argument that the Pythagoreans had a special interest in seeing to it that no harm or injustice was done to patients (by the patients themselves, especially) owing to their adoption of immoderate or unhealthy eating habits. I do not wish to deny that for the Pythagoreans dietary practices were closely wedded to purificatory practices carrying religious and moral overtones. But, even so, the imperative to keep the patient from harm or injustice is very similar to another Hippocratic imperative found in *Epidemics I*, which says: "As to diseases, make a habit of two things – to help, or at least to do no harm."[27] Surely no conscientious Hippocratic physician would have quarreled with this basic teaching. As Edelstein himself admits, "It is the goal of all good craftsmen to seek the best for the object with which the craftsman is concerned."[28] Hence I must also differ with Edelstein's tendency to see this injunction to protect the patient as referring to Pythagorean dietetics alone. It seems to me better understood as a pledge to protect the patient from harm and injustice in *other* areas of medical treatment as well, not just dietetics. Read in this equally plausible way, Edelstein's argument loses force.

In addition, consider (v) above, the imperative not to have sexual intercourse with the patient or any member of the patient's household — freeborn or slave, female or male. Here, too, Edelstein fails to establish that this unconditional prohibition expresses uniquely Pythagorean values or is understandable *only* in connection with Pythagorean doctrine. I do think he makes a good case for the view that the Pythagoreans probably judged sexual relations in terms of just or unjust conduct. There is little doubt that they condemned adultery and sodomy, and that they had high praise for marital fidelity. Moreover, in their performance of moral duties the Pythagoreans taught that one should not discriminate between freeborn persons and slaves.[29]

Nevertheless, DeVogel is right in resisting Edelstein's characteristically exclusive claims here on behalf of Pythagoreanism.[30] Did the Pythagoreans alone oppose adultery, sodomy, and sexual injustices, among all the Greeks? It appears not. The Stoics, for example, were among those who shared in the condemnation of adultery.[31] It is equally questionable, as Edelstein seems to imply, that both homosexuality *and* adultery were opposed by no other fourth century philosophers except the Pythagoreans. Plato, for one, condemns both in his *Laws*.[32] In addition, one discovers in the Hippocratic treatise *The Physician*, which primarily addresses matters of proper medical etiquette, the following passage:

In every social relation he will be fair, for fairness must be of great service. The intimacy also between physician and *patient* is close. Patients in fact put themselves into the hands of their physician, and at every moment he meets women, maidens, and possessions very precious indeed. So towards all these self-control must be used. Such then should the physician be, both in body and in soul.[33]

Edelstein is of course aware of this preceding Hippocratic parallel. But he labors to discredit it as different in kind from (v) because (a) the preceding passage speaks only of the avoidance of injustice, whereas (v) includes the avoidance of "mischief" (consult the Oath (P6), p. 69, above); and because (b) the preceding passage, unlike (v), literally refers only to women (not men or slaves).

I do not wish to minimize these differences. But it seems fair to conclude that we are faced here with at most a difference in the stated degree of the prohibition only. In fact, the preceding passage also includes such introductory phrases as "in *every* social relation" and refers to "patients" in general —

without qualification. Therefore, it is a bit hasty to regard it as intended for women or the freeborn alone, as Edelstein would have us think. Thus, here again, the singularity of his Pythagorean thesis does not fit the facts.

On the matter of protecting the privacy of patients, it is of no small significance that the Oath includes the imperative (vi) that the physician not gossip about his patients. He is sworn to silence, both at home and abroad, about information he acquires " . . . in the course of treatment or even outside of the treatment in regard to the life of men . . . " (P7).

In this connection, Edelstein reminds us that the Pythagoreans greatly valued reticence. Silence was part of the Pythagorean way of life. The Pythagoreans were not in the habit of imparting their knowledge to others; they did not divulge everything to everybody. Moreover, for their secrecy and frugality they were sometimes made the butt of ridicule in ancient comedy.[34] Presumably the Pythagoreans considered reticence a high moral duty, not merely a matter of convention. Also, they apparently held that the scientist or aspiring pupil ought to acquire a quiet, attentive demeanor in order to attain a clear state of mental concentration receptive to fresh discoveries. To be able to consistently demonstrate such reticence was considered a sure sign of moral self-control.[35]

Here I must concur with Edelstein that P7 is also imbued with a similarly strict moral tone. It is an unconditional pledge. It is strongly implied that to violate this pledge would amount to a serious violation of duty. Moreover, it is not hard to imagine that there were those in Greek culture, as in our own, who sought to secure some advantage for themselves by learning from a physician the details of a patient's physical or mental condition, or about other such personal matters. If so, the author of this Oath deemed such an indiscretion "shameful." Thus the Oath, in (vi) as elsewhere, establishes in no uncertain terms the primary moral duties and responsibilities that define a good medical relationship between doctor and patient.

Against Edelstein, Kudlien has argued that this Pythagorean tendency to value silence and eschew gossip (which seems mirrored in the Oath) was not unique.

First, the command to observe the *aporrheta* [state secrets] is very common in other Greek oath-formulations as sworn by citizens and officials. Second, it is an old command of Greek popular ethics (or polis ethics) to avoid the uttering of *aporrheta*, a command that may have been still current in later epochs, at least in professional and political (civic) ethics if this latter were enforced in the rigid, and traditionally solemn, form of the oath.[36]

However, Kudlien's evidence in support of these claims is tenuous. He cites, in particular, some of the proverbial passages attributed to the so-called seven wise men of antiquity. Also, it is far from clear whether such pledges to secrecy as he cites, especially those associated with the popular (i.e., non-philosophical) ethics common to the Hellenistic area, were much more than guides to social etiquette. In contrast, the pledge to remain silent in P7 of the Oath is absolute and carries unmistakable moral overtones of a sort unmatched by Kudlien's allegedly parallel examples. Hence I do not take Kudlien's assertions here as by any means proved.

ABORTION AND EUTHANASIA

Finally, we come to the all-important imperative against aiding a patient either in euthanasia or abortion ((ii) and (iii), respectively). Edelstein again attempts to narrow the alternatives in such a way that Pythagoreanism emerges as the best explanation of the source of these two moral constraints. In addition, Edelstein argues that these expressed contraints do not agree with either the dominant moral outlook or actual practices of the ancient physicians of Greece and Rome. That is, not until Christian values became dominant during the last century-and-a-half of the Roman Empire (particularly the Judeo-Christian value of respect for human life), did the prohibitions against abortion and euthanasia gain support in ancient medical circles. Therefore, at best, the prohibitions in question long represented a small segment of Greek opinion, Edelstein contends.

Is Edelstein right in identifying the Pythagoreans as the likely authors of these two prohibitions? It will be helpful to directly consider the premises of his argument.

Ancient jurisdiction did not discriminate against suicide; it did not attach any disgrace to it, provided that there was sufficient reason for such an act. And self-murder as a relief from illness was regarded as justifiable, so much so that in some states it was an institution duly legalized by the authorities. Nor did Greek or Roman law protect the unborn child. If, in certain cities, abortion was prosecuted, it was because the father's right to his offspring had been violated by the mother's action. Ancient religion [i.e., Homeric religion] did not proscribe suicide. It did not know of any eternal punishment for those who voluntarily ended their lives. Likewise it remained indifferent to foeticide. Its tenets did not include the dogma of an immortal soul for which men must render account to their creator. Law and religion then left the physician free to do whatever seemed best for him.

From all these considerations it follows that a specific philosophical conviction must have dictated the rules laid down in the Oath. Is it possible to determine this particular philosophy? To take the problem of suicide first: Platonists, Cynics, and Stoics can be eliminated at once. They held suicide permissible for the diseased .... Aristotle, on the

other hand, ... and Epicurus [were opposed to suicide but this] ... did not involve
moral censure. If men decided to take their lives, they were within their rights as sovereign masters of themselves.

Pythagoreanism, then, remains the only philosophical dogma that can possibly
account for the attitude advocated in the Hippocratic Oath. For indeed, of all the Greek
thinkers the Pythagoreans alone outlawed suicide and did so without qualification . . . .
Most of the Greek philosophers even commended abortion. For Plato foeticide is one
of the regular institutions of the ideal state . . . . Aristotle reckons abortion the best
procedure to keep the population within the limits which he considers essential for a
well-ordered community.[37]

What I shall try to show in the remainder of this section is that Edelstein's
exclusive singling out of the Pythagoreans as the *only* plausible source for the
prohibitions against abortion and euthanasia is not fully justified. The Pythagorean influence cannot be ruled out, I shall argue. But neither can it be fairly
seen as the only likely one operating here. Hence I shall postpone until Part
Three a detailed examination of selected philosophical perspectives and
ancient medical practices associated with abortion and euthanasia in the
Greco-Roman era. My present task is to clarify in a preliminary way the likely
meaning of these twin prohibitions for those who may have taken the Oath.
For the time being, then, most of the philosophical attributions concerning
which Edelstein was just quoted shall be set aside pending further investigation in the two chapters which follow.

It will be recalled that the paragraph in question (P4) is this:

*Ou doso de oude pharmakon oudeni aitetheis thanasimon oude hyphegesomai xymbou-
lien toiende; homoios de oude gynaiki pesson phthorion doso. Agnos de kai hosios
diatereso bion emon kai technen emen.*

I will neither give a deadly drug to anybody if asked for it, nor will I make a suggestion
to this effect. Similarly I will not give to a woman an abortive remedy. In purity and
holiness I will guard my life and my art.

As to the first sentence, it becomes tempting to ask whether we have here
something more than a prohibition against assisting in euthanasia or suicide.
That is, can this sentence be plausibly understood in a somewhat wider sense
to formally prohibit, in addition, the ancient physician from involvement in
any sort of murder plot as the supplier of effective poisons? Edelstein denies
this. But in so doing, I think he is over-simplifying the full meaning of this
constraint. He needs to deny this wider meaning. If it is admitted that other,
non-Pythagorean elements may have shaped this proviso of the Oath, it
becomes increasingly difficult for him to make his exclusively Pythagorean
thesis stick.

I hold that the author of this proviso may have intended at least a tacit reminder to the physician that under no circumstances does his profession morally permit him to be an accomplice to murder. There are several considerations that render this wider meaning acceptable. First, it is a documented fact that in the Greco-Roman era it was popularly believed that the physician who is skillful at curing was, in virtue of this knowledge, also peculiarly well suited to know how to kill. One can find Plato echoing this suspicion in his *Republic*. In Book One, Socrates asks Polemarchus: "Who then is the most able when they are ill to benefit friends and *harm enemies* in respect to disease and health?" To which Polemarchus correctly replies, "The physician." [38]

Moreover, in Book Eleven of his *Laws*, Plato specifies that any physician who is found guilty of willfully attempting to poison another (whether the attempt is fatal to the victim or not), " . . . *must* be punished by death."[39] It is worth noting that, in contrast, the *Laws* dictates that any *layman* found guilty of the identical crime need not face capital punishment. "The court is to decide the proper penalty or *fine* to be inflicted in his case."[40] This passage shows, I think, the special contempt (if not veiled popular prejudice) that Plato felt toward the Greek doctor who plied his trade in vicious ways. Such a person, by dint of the special knowledge embodied in his craft, in some sense knows (in the way a layman ordinarily cannot) the outcome and special effects of drugs on the body. Hence the doctor is all the more responsible for his crime: his crime is at once all the more vicious.

Edelstein's attempt to discredit the possibility that Plato's *Laws* may here share a common element of disapproval, paralleled in this wider reading of the Oath's stricture against poisoning, is hardly convincing. Edelstein says that the stricture under discussion would then represent a useless or absurd duplication of existing Greek laws.[41] But suppose there was just such a duplication. If so, it is hard to see why this would therefore prove useless. Such a reinforcement of the aforementioned Hippocratic value to "at least do no harm"[42] may plausibly have stood as a very useful moral supplement to the homicide laws of the Greek city-states. In fact, ancient oaths sometimes did incorporate various features of contemporary legal or religious norms. So it would be overhasty to exclude the influence of popular suspicion against the potential power of physicians to do harm in either medical or non-medical contexts. Thus we must not rule out the possibility that the Oath's proviso against poisoning represents in the broadest sense a conservative reaction to such suspicions and popular prejudices.[43] The author of the Oath may therefore not have been a Pythagorean, even though his *prima facie* respect for human life would have admittedly proved congenial to Pythagorean physicians and philosophers.

To this I must add that there is no evidence that it was illegal under the laws of the Greek city-states for physicians to supply a person who wished to commit suicide with a suitable drug. Nor during the Roman Empire was this a criminal act.[44] As Plato's *Laws* suggest, it *was* a criminal act for any person, physician or layman, to poison another citizen in a murderous plot. Why, then, should such an obviously wrong deed as murder by poisoning receive mention in a doctor's code of ethics? Again, it is worth emphasizing that times were different then. Anyone could practice medicine, and many who did were quacks and charlatans. Money and greed were powerful motives for complicity in murder. As P4 itself presupposes, ancient physicians themselves dispensed pharmaceuticals and so were in the position to sell the necessary poisons in order to enrich themselves.

In addition, it is very much in doubt whether the Greeks possessed sufficient knowledge of pathology and post-mortem medical diagnosis to be able to detect death by poisoning, or deduce by chemical analysis the likely person or persons who slipped the deadly compound to the victim.[45] This ignorance of forensic and post-mortem medicine may have redoubled the temptation of unscrupulous doctors to peddle their knowledge and their drugs for ill-gotten gain.[46] If so, then this supplies all the more reason to think that the prevention of murder, as well as suicide or euthanasia, may have been relevantly addressed in this single and elusive sentence of the Oath.

As for the next sentence of P4, which, as it is rendered here absolutely prohibits abortion, some may wish to challenge whether Edelstein's translation of this sentence is correct.[47] Personally, I am convinced that it is. This, despite the fact that his rendering of *pesson phthorion* as "an abortive remedy" goes significantly beyond what the Greek literally says. Literally translated, one gets "a poisonous pessary." The point is that pessaries (vaginal suppositories) were but one of several methods then used for inducing abortion. Given this, a case can be made that, at most, the Hippocratic Oath prohibited just one type of abortion, not all — as it has commonly been taken to do.[48] But is such a literal translation sound? Since so much of our ensuing investigation depends on getting this passage straight, it will be necessary to consult the pertinent internal and external evidence.

What counts most against what I shall call the literalist interpretation, in view of the internal evidence of the Oath itself, is the parallel logic which is implied in the first two sentences of P4. For example, the Greek word *homoios* means "in like manner," or "similarly."[49] *Homoios* introduces the restriction in question concerning abortion immediately following a sentence which all agree unconditionally prohibits the Oath-taker from assisting in acts of active euthanasia (whatever else that sentence may accomplish). Consider: does it make good sense to suppose that the author of the Oath would introduce his second major restriction by employing the strong adverbial modifier "similarly," if, as the literalist insists, the author intended merely a conditional prohibition here? A prohibition which *ex hypothesi* singled out and excluded just one of the several abortion methods at the physician's disposal?

The point is that any literalist translation of this sentence faces the burden of explaining on what grounds partisans of the Oath (so construed) held abortifacient pessaries to be morally objectionable but not such related abortion methods as the manual puncturing of the womb; the drinking of allegedly abortifacient potions; or forms of violent bodily exercise. In fact, external evidence drawn from contemporary and subsequent medical sources in antiquity suggest that these alternative methods were widely known to physicians, midwives, and others.[50] What, then, made these remaining kinds of abortion methods morally unobjectionable in the Oath-taker's mind? For did these not aim at the identical result, i.e., the willful termination of the life of the fetus?

One possible answer is that abortifacient pessaries were capable of being so powerfully concocted that they sometimes led not only to the intended death of the fetus, but to the unintended death of the pregnant woman (by infection?).[51] But, if so, how odd that there is scarcely a whisper of evidence to this effect in the ancient medical writings. One might well expect to find some mention of such a tragic outcome in the writings of the first century A.D. gynecologist Soranus, if anywhere.[52] After all, he warns of many other treatment hazards which he would have his peers and his patients eschew.

Yet, notwithstanding these difficulties, the literalist still has at least one seemingly powerful rejoinder. He can remind us that *pesson phthorion* unequivocally means "a poisonous pessary." The text does not say "an abortive remedy," as Edelstein contends. Nor, by any reasonable stretch of the imagination, can it be legitimately so read. Granted, he continues, a literal reading of *pesson phthorion* may cause us to confront a whole web of loose ends that betray a coherent grasp of the ancient problem of abortion. But let us not invent a reading of the text in a vain effort to rescue our preconceived notions about what Hippocratic medical ethics does or does not permit.

This last-ditch literalist defense is not persuasive. To insist that the text can be reasonably understood only in the literalist cast on the grounds that no other interpretation is eligible, is manifestly question-begging. Alternatively, I speculate that by the time physicians were enlisted for their help in procuring an abortion, the patient (in conjunction with compliant midwives, old family nurses, or other confidentes) usually had tried in vain to dislodge the fetus using mechanical methods which were generally known to experienced women at that time. The prepared pessary was probably the most common abortifacient remedy dispensed by physicians; and for it there were recipes of variable effectiveness and safety. Typically, going to a physician for an abortion was the last resort: the proper preparation and insertion of the pessary required something more than the ordinary layman's knowledge of first aid.

Seen against this background, I further conjecture that the Oath's literal prohibition against giving poisonous pessaries had the symbolic moral force of excluding the physician from *all other* modes of inducing abortion. Mention of the deadly pessary functioned linguistically to signal opposition to the entire range of abortive remedies in roughly the same way that, today, a person's explicit statement that he opposes the hanging of criminals is usually taken to mean that he opposes all forms of capital punishment. If so, the poisonous pessary prohibited in P4 functioned linguistically to announce an absolute restriction on the entire class of abortion techniques. Hence, because he tacitly embraced a version of the respect for human life principle,[53] any one of these life-destroying maneuvers was morally repugnant to the Oath-taker, no less so than the termination of human life involved at the other end of the spectrum in the equally prohibited act of active euthanasia. I conclude that no alternative translation of this passage coheres so well with the fuller text of the Oath, nor explains measurably more so concisely.

As for the next sentence of P4, which prohibits dispensing drugs for the purpose of inducing abortion, it is a questionable premise whether this prohibition represents, as Edelstein holds, an *exclusively* Pythagorean value. I do not wish to dispute that the Pythagoreans saw the fetus as an animate human life unconditionally worthy of preservation from the moment of conception. It is also undeniable that children were greatly valued by the Pythagoreans on religious, and not merely on socio-economic grounds. That is, it was considered by them each man's duty to beget children so as to leave behind at death another earthly worshipper of the gods. Even so, here doubts arise about whether the Pythagoreans *alone* resisted abortion, as Edelstein needs to demonstrate. William Frankena, for one, has questioned this assumption, though he fails to supply any damaging counter-examples.[54]

More importantly, Kudlien has argued that there is some evidence that non-Pythagorean Hellenistic cult laws also viewed miscarriage or abortion with distaste. These bloody events were seen as possible sources of ritual uncleanliness and impurity. As such, they were condemned as cult taboos among some of the very religious.[55] If Kudlien is right, as I am inclined to think, then Edelstein's thesis is again weakened. As I have argued all along, it is too narrowly conceived. In the present case, it is equally possible, I submit, to understand the origins of the Oath's constraint against abortion as the product of a religiously-minded person with a particular fondness for cult rules and institutions. Such an understanding also fits in well with the last sentence of P4: "In purity and holiness I will guard my life and my art." The Greek words *hagnos kai hosios* (in purity and holiness) do signal, as Edelstein correctly argues, a distinctly devout attitude on the part of the author. But versions of this phrase were generally used in cult oaths and rituals — a fact Edelstein does not share with his readers.

THE OATH'S DATE AND ORIGINS RECONSIDERED

My final departure from Edelstein's account concerns his attempt to locate the Oath specifically in the fourth century B.C. In view of the dearth of evidence at our disposal, I do not myself claim to know the approximate date of the Oath's authorship. If pressed, I would say it was probably written between the sixth and third centuries B.C. Already I have shown that Nittis' attempt to date it within a six-month span in the year 421 B.C., in Athens, is over-determined and unpersuasive. And, for different reasons, I find Edelstein's judgment equally unpersuasive. Edelstein's argument for narrowing the Oath to the fourth century turns on a false dilemma. Suppose, for the sake of argument, that the Oath is entirely of Pythagorean origins. Even so, it is overhasty to infer, as he does, that owing to our ignorance of Pythagorean developments in the two prior centuries, the Oath is therefore not a very likely product of the sixth or fifth centuries. It is also overhasty to exclude the third century just because the Pythagorean movement held less sway over Greek and Roman intellectuals during that later Hellenistic period. As DeVogel has aptly pointed out:

The [Pythagorean] School flourished during the 6th and the first half of the 5th centuries, and if there is still a tradition of Pythagorean forms of behavior during the 4th century, this is due to the things that were handed down by previous generations. The Pythagorean principles . . . — not to kill, not to dispose of one's own or someone else's life, the reverence for life even if it is not yet born, the avoidance of surgery even in cases where others commonly used it, the reverence for the human person — all these were doubtless of ancient standing.[56]

All things considered, then, the conclusion I am willing to support regarding the Oath's origins, date, and purpose is this: Edelstein is at best half-right about the date of the Oath. It may have been written in the fourth century. But I see no reason why it could not have originated a century later, or during the two previous centuries, for that matter.

As to its origins, the Oath does contain moral and religious elements, in its covenant and ethical code especially, that are fully compatible with a Pythagorean influence. This much I grant. But I have also demonstrated that it is likely that *additional influences* — from Greek religious cults, homicide laws, and popular ethics — may have affected the form and content of the Oath as well.

As to its purpose, there is probably some substance to Edelstein's suggestion that the Oath served as the tool of a reform movement in Greek medicine that broke away from the more dominant ethical values and medical practices of Classical times. But it is far from certain that the Oath was therefore an esoteric document, though I do not rule this out. Indeed, at least some of the evidence introduced in this section implies that the Oath may have reflected a broader and more diverse range of popular, religious, and social sentiments than one might have at first supposed. In any case, it does seem likely that at best only a small number of physicians in antiquity actually swore and abided by the Oath's moral directives. Those that did were nonconformists. Whether they acted as individuals or were part of a secret medical fraternity devoted to a blend of Pythagorean and other ideals is hard to say. Personally, I tend to think that the Oath may have been used by some ancient medical fraternities or guilds. At least, nothing we have discovered so far excludes this possibility.

MEDICAL ETIQUETTE

To complete the present account of the rise of Greek medical ethics, we turn now to explore four additional Hippocratic works which promise to shed more light on the values and conduct of the ancient physician. Among the things these four Hellenistic works have in common is that, while their exact origins are unknown, they all specify in some detail the good manners that the ideal Hippocratic physician should display in the practice of his craft. Since they address such topics as professional conduct between physicians, bedside manners, proper decorum, and handling difficult patients, these four works are traditionally counted as the basic Hippocratic guide to sound medical etiquette. In contrast, the Hippocratic Oath may be characterized as a more substantive and unified work. As the preceding section suggests, its ethical code appears to be founded on moral principles of the highest sort. I shall show in what follows that, while the four works in question do primarily involve lessons in etiquette, they also manifest a sincere concern for the patient's welfare. This concern goes beyond the ancient physician's understandable inclination to promote his own professional reputation and secure his own pecuniary self-interest by the cultivation of patronizing bedside manners.

The Greek word for etiquette, *euschemosyne*, literally means being graceful, elegant, manifesting good form or bearing. This word can be usefully compared with the Greek *ethika*, meaning of or for morals.[57] For what is done for morals or out of moral concern was ordinarily taken by the Greeks to involve something more like the performance of duties (to oneself or others) than the mere display of good form, though it may have incidentally involved that, too. This subtle difference between ethics and etiquette is confirmed by observing that *euschemosyne*, not *ethika*, carried with it a derogatory sense meaning good in outward show, i.e., specious. I begin by calling attention to the Greek grammar of these two concepts because some ambiguity exists in recent discussions of ancient professional ethics on just what sets medical ethics apart from medical etiquette. The tendency has been to conflate any distinction between ethics and etiquette, or ignore it altogether. The safest line of demarcation, and one fully borne out by the four Hippocratic texts about to be examined, is that, in general, a concern for etiquette and for doing the right thing morally *do very often overlap*. But etiquette concentrates primarily on the outward show of correct behavior; thus it is usually associated with courtesy and what is popularly termed good breeding. In contrast, moral prescriptions are primarily concerned with making the right choice and doing the right thing based on some favored principles of human duty; thus moral prescriptions ultimately transcend a mere concern for outward show and ceremony.

In addition, it is worth recalling that for the Greek physician the rules of etiquette (as well as ethics) were followed not through fear of civil punishment — for there was no regulation of medicine. Ideally, the physician was motivated to acquit himself well partly out of a love for his craft. Also he was motivated by a somewhat more egoistic concern to acquire and preserve a good reputation. For his reputation was the only professional credential the ancient physician had. Rules of etiquette, as such, had nothing to do with the application of categorical imperatives. One gets the impression that, at most, they were *rules of thumb* that implied "should" or "ought," never the

unequivocal "must." Yet etiquette was nonetheless of considerable impc
tance to the successful practice of medicine. For, in the absence of medic
licensure, the financial success of one's practice depended almost entire
on the patients' perceptions of the physician (regardless of what medical sk
the physician, in fact, possessed).

Having considered all this, let us begin with the oldest and shortest of tl
manuals of Hippocratic etiquette, the very brief tract entitled *Law*. My inte:
tion throughout is to let these excerpts speak for themselves, except for m
mention of a few background details on each piece.

Law was known to the compiler and physician Erotian (fl. 50 A.D.) bι
is mentioned by no other ancient author. Both Jones and Edelstein date
to between the fifth and fourth century. Whether it is a fragment or a con
plete work (it runs only about 300 words), no one is sure. Jones suggests thε
its name derives from the fact that it lays down the law for a good medicε
education. He further fancies that *Law* may have been a short address by th
head of a medical school to graduating physicians.[58] At any rate, our intere:
in it here is that it begins by complaining of the lack of any state regulatio
over those who practiced medicine. Incidentally, the lament that the art o
medicine is a noble profession but much maligned owing to the quacks in it
ranks, is echoed by physicians throughout antiquity.

Medicine is the most distinguished of all the arts, but through the ignorance of thosι
who practice it, and of those who casually judge such practitioners, it is now of all thι
arts by far the least esteemed. The chief reason for this error seems to me to be this
medicine is the only art which our states have made subject to no penalty save that o
dishonor, and dishonor does not wound those who are compacted of it. Such men in fac
are very like the supernumeraries in tragedies. Just as these have the appearance, dress
and mask of an actor without being actors, so too with physicians; many are physician:
by repute, very few are such in reality.[59]

The author goes on to sketch what natural abilities and type of education arε
required for becoming a competent physician. Moreover, consistent with the
author's joint concern for advancing his profession's interests, as well a‹
meeting the genuine needs of patients, he lectures the reader to acquire a
real knowledge of medicine — not just a good outward show. This must be
done " . . . before we travel from city to city, and win the reputation of being
physicians not only in word, but also in deed."[60]

A second, somewhat later work, is entitled *The Physician*. It probably
dates from about 350–300 B.C., and like *Law* and the remaining *Precepts*
and *Decorum*, it was probably intended as prudent advice for young doctors.
Unlike the other three tracts, however, this piece goes on in its later chapters

to address methods of bandaging and surgery. The book ends by advising students of medicine to try to find work caring for mercenary troops in order to acquire valuable experience in the practice of surgery.[61] All this is consistent with the fact that war-related injuries supplied a prime source for the study of human anatomy and physiology in the ancient world. Galen himself was the former physician of gladiators, and he alluded to the valuable use of combatants for the gain of medical knowledge.[62]

It will be recalled that I have already quoted from the first chapter of *The Physician* to demonstrate the author's general concern for fairness in social relations along with his stated opposition to the sexual exploitation of patients (above, p. 79). Yet there are passages of another sort to be found in that same chapter which reveal a more personal, egoistic concern for the physician's reputation alone. These directives have to do with such comparatively mundane matters as the proper deportment, dress, and demeanor of the socially correct doctor.

The dignity of a physician requires that he should look healthy, and as plump as nature intended him to be; for the common crowd consider those who are not of this excellent bodily condition to be unable to take care of others. Then he must be clean in person, well dressed, and anointed with sweet smelling unguents that are not in any way suspicious. This, in fact, is pleasing to patients. The prudent man must also be careful of certain moral considerations – not only to be silent, but also of a great regularity in life, since thereby his reputation will be greatly enhanced; .... In appearance, let him be of a serious but not harsh countenance; for harshness is taken to mean arrogance and unkindliness, while a man of uncontrolled laughter and excessive gaiety is considered vulgar, and vulgarity especially must be avoided.[63]

Turning finally to *Precepts* and *Decorum*, which are longer works and probably date to the first century B.C., we are still in the dark as to their origin. Both appear to evidence a possible Stoic influence. But, unfortunately, attempts to convincingly link any of the four manuscripts to specific philosophical schools have been fraught with difficulty.[64] They are too widely eclectic and so do not reveal a clear moral ideology. What's more, it is probably misleading to infer that these works were widely read and faithfully followed by the run-of-the-mill Greek or Roman doctor. Collectively, these works represent a considerable commitment to the best medicine could then reasonably offer to patients in need. Any one at all could practice medicine. Many apparently unsavory characters did so with the littlest regard for the welfare of their patients, let alone the nuances of etiquette. When their reputations caught up with them, they traveled on.

Notwithstanding these considerations, *Precepts* and *Decorum* are philosophically important along with the others. They serve to demonstrate the higher road that Greco-Roman medicine was capable of taking. They also demonstrate an understandable *tension* between furthering the self-interests of the practitioner, on the one hand, and providing sound medical care aimed at the best interests of patients, on the other. This mediation between egoistic and altruistic interests is evident in the following excerpt from *Precepts*. Dealing essentially with fee-setting, it admonishes the physician to consider the wealth of his patients in seeking the fullest possible payment for services rendered; at the same time, the physician is reminded not to practice without compassion and attention to the needy.

I urge you not be too unkind, but to consider carefully your patient's superabundance or means. Some times give your services for nothing, calling to mind a previous benefaction or present satisfaction. *And if there be an opportunity of serving one who is a stranger in financial straits, give full assistance to all such.* For where there is *love of man*, there is also love of the art. For some patients, though conscious that their condition is perilous, recover their health simply through their contentment with the goodness of the physician. And it is well to superintend the sick to make them well, to care for the healthy to keep them well, *but also to care for one's own self*, so as to observe what is seemly.[65]

In *Decorum*, as in the other three works on etiquette, disapprobation is expressed against the vulgar and untutored quacks who masquerade as physicians and so threaten to spoil the reputations of all. Characteristic of these works on etiquette, every effort is made to remind the young physician of the grand ideals to which medicine at its best aspires. Here, as elsewhere, there is no explicit mention of the Hippocratic Oath. Nor is there a sustained discussion of moral principles. There is, however, what appears to be a utilitarian emphasis throughout which attempts to balance the doctor's needs with the needs of his patient. These next four passages may suffice to give the flavor of the author's three central aims. These are: (1) to proclaim the exemplary intellectual and moral ideals of the medical art; (2) to provide for his patients' best interests; and (3) to secure for himself the best reputation and greatest financial success he can.

For a physician who is a lover of wisdom is the equal of a god. Between wisdom and medicine there is no gulf fixed; in fact, medicine possesses all the qualities that make for wisdom. It has disinterestedness, shamefastness, modesty, reserve, sound opinion, judgment, quiet pugnacity, purity, sententious speech, knowledge of the things good and necessary for life, selling of that which cleanses, freedom from superstition, pre-excellence divine.[66]

. . . .

Make frequent visits; be especially careful in your examinations, counteracting the things wherein you have been deceived at the changes . . . .[67]

. . . .

Keep watch also on the faults of the patients, which often make them lie about the taking of things prescribed. For through not taking disagreeable drinks, purgative or other, they sometimes die. What they have done never results in a confession, but the blame is thrown upon the physician.[68]

. . . .

On entering [the patient's room] bear in mind your manner of sitting, reserve, arrangement of dress, decisive utterance, brevity of speech, composure, bedside manner, care, replies to objections, calm self-control to meet the troubles that occur, rebuke of disturbance, readiness to do what has to be done.[69]

It is worth noting that to the extent that *Law, The Physician, Precepts*, and *Decorum* do express a commitment to protect the best interests of the patient, these works may be said to be logically consistent with the patient-benefiting ethic found in the ethical code of the Hippocratic Oath. Yet there is a marked *difference in emphasis*. The ethical code of the Oath is manifestly bereft of a concern for mere manners and outward deportment. It speaks of " . . . remaining free of all *intentional* injustice."[70] In general, the Oath expresses a pledge to define and preserve the right conduct of the physician-patient relationship, based ultimately on ethical principles and concern for the intentions of the *inward*-reflecting person. Anything to be gained by *mere outward show* and the ready cultivation of clever manners is at best a secondary consideration to the Oath-taker. Strictly speaking, in swearing the Oath he gives no notice to these things.

A decidedly different emphasis, however, pervades the four Hippocratic works just discussed. The guiding and persistent interest here is unabashedly that of doctors who would choose to live prudently. The tone and content of these writings is generally more practical than the Oath. Yet it would be harsh, I think, to judge such repeated concerns for proper bedside manners and deportment out of the context of the ancient medical marketplace. There, cultivating a sound medical reputation in the absence of any system of licensure was especially critical. Notwithstanding this fact, I have shown that the welfare of the patient is never really completely subordinated to the physician's more selfish interests. Nor is it ignored. Rather, the patient is again and again taken into consideration, and, very often, not without evidence of sustained human compassion.

SUMMARY

If we return, now, to reconsider the principal questions with which Part Two began, the evidence and analysis of the preceding pages can be summarized as follows.

The Hippocratic Oath is a unique medical ethical document from at least two fundamental perspectives: (1) the pan-Mediterranean perspective, and (2) the pan-Hellenic perspective. Seen in the broader context of the pan-Mediterranean world, the Oath appears to stand without parallel or precedent in Egypt, Assyria, Babylonia, and Persia. If these neighboring and more ancient cultures produced an oath, prayer, treatise, or philosophical-religious document of any sort devoted exclusively to the ethics of medicine, there is no record of this in their literature, art, or science.

As for their laws, the Babylonians may be credited as the first civilization on record to create laws designed, among other things, to regulate medical practice. The Babylonian Code of Hammurabi set forth stiff fines against malpractice more than a thousand years before the Hippocratic Oath came into being. But these laws were confined to surgical procedures alone. Also, patients did not receive equal protection under the laws of Hammurabi: the degree of legal protection one received against medical malpractice varied in proportion to the social status of the patient. Therefore only a fraction of Babylonian medicine was brought under the legal jurisdiction of the state, and individuals seeking surgical treatments, at least, were apparently not deemed to be of equal worth.

This stands in marked contrast to the Hippocratic Oath (and, also, it would seem, the later Hippocratic treatises concerned with etiquette). In the Oath, a single standard of medical care is proposed for all patients, protecting even the slave. In addition, the Oath is probably not a legal document. It was not created by any legislature, nor was it made legally binding on any Greek physician throughout antiquity, so far as we know. How the Oath specifically came into existence is a puzzle. But one thing seems beyond doubt. It arose largely as a response by its author or authors to the free and unrestrained marketplace which Greek medicine enjoyed. It was very likely the result of a felt need by a handful of Greek physicians to fill a void which had been created by the absence of any legal constraints on the practice of medicine. Thus, the Oath was *voluntarily* obeyed by those few who chose to adopt it; it emerged out of the precondition of a relatively high degree of personal and professional freedom. None of this can be said of the state-imposed regulations of medicine found among the neighboring Babylonians, Assyrians, and Egyptians.

Furthermore, while there is evidence to suggest that, on either legal or religious grounds, abortion was probably once opposed by the Assyrians and Persians, if not the Egyptians, we are not privy to the moral reasoning behind this opposition. As to the moral, legal, or religious stands these cultures took against euthanasia, it *is* known that in Egypt and Assyro-Babylonia, the act of suicide was thought to cut one off from the gods. But any evidence indicating their specific condemnation of euthanasia is wanting.

Quite the opposite is true when one considers the Hippocratic Oath. Both abortion and euthanasia are unequivocally opposed in P4. And to the extent that this opposition may have been found on something akin to a respect for human life ethic, at least some hint of the moral reasoning behind these powerful prohibitions may be secured. Clarifying the premises of this reasoning is part of the task that awaits us in Part Three.

Equally significant, even though the ancient cultures of Egypt, Assyria, Babylonia, and Persia gave at least *some* thought to the proper conduct of their physician-priests, no evidence suggests that their medical ethical values in any way influenced either the form or content of the Hippocratic Oath. The same may be said for the Hippocratic treatises on medical etiquette.

On the other hand, when the Oath is seen in the narrower context of the pan-Hellenic world alone, its uniqueness becomes apparent in at least two ways. First, the Hippocratic Oath stands as the only Greek medical ethical document of its kind. If there were other such medical oaths, we lack any evidence of their existence. One is left with the distinct impression that craft-related documents like this were in any case quite rare.[71] Second, although the Oath was not legally binding, it is fair to say that the power over men's minds which it exercised derived mainly from the practitioner's sense of moral duty and personal conscience. Whether or not the Oath is entirely a Pythagorean manifesto, as Edelstein has argued, or, as I think, only partly so, the primary duties it involves can be reduced to two fundamental kinds: First and foremost, the duty to protect the patient from harm and to rid him of disease. This is a duty to others. Second, the duty to preserve and promote the good reputation of the medical craft by adopting practices designed to insure the integrity and competency of its members. This is primarily a duty to oneself (though it also affects others). Hence, the Oath is unique in formally defining for some Greeks the proper conduct of the physician-patient relationship. It goes a long way toward putting this relationship on a *prima facie* moral and religious footing. Even so, as I have implied, the Oath may have been influenced by more prudential considerations as well. The latter may include the desire to avoid public controversy (in regard to surgery, abortion, and euthanasia, especially), and the desire to win the lasting admiration of both gods and men (P8). But even these aims, taken in context, were considered praiseworthy. For, to quote from the *Precepts*: "Where there is love of man, there is also love of the Art."[72] Perhaps the same may be said for the effects of love of the health-related gods like Apollo, Asclepius, and the latter's two daughters Hygieia and Panaceia, before whom the Oath is specifically sworn. For did they not likely motivate the Oath-taker to higher aspirations for both himself and his patients, too?

Lastly, if the riddle of the Oath's opposition to the widespread ancient practices of abortion and euthanasia has not been altogether solved, then perhaps it may now be dissolved. For if the Oath is, as I have argued, a possibly esoteric document partly informed by Pythagorean tendencies, and if, in addition, it does represent some sort of reform movement in Greek medical practice, then it is plausible to infer that its adherents constituted a relatively small group of physicians. This does not necessarily mean that those who refused to join them were regarded by the Oath's adherents to be practicing bad medicine. But it does mean that these dissenters, who were in the majority, might have been influenced by somewhat different ethical values. Hence in Part Three, to which we now turn, I shall attempt to isolate and analyze just what some of these competing philosophical perspectives on abortion and euthanasia may have been.

PART THREE

ABORTION AND EUTHANASIA

CHAPTER SIX

THE PROBLEM OF ABORTION

We come now to the third part of this study into the origins of Western medical ethics. Having already investigated the social setting of ancient medicine as well as the rise of European medical ethics with Hippocrates and his successors, we are at last in the best position to explore the thinking of the ancients on abortion and euthanasia. Let us, then, see what the leading philosophers and physicians had to say about these topics. In so doing, let us also be careful to look at their related thoughts on infanticide and suicide. Our final task involves clarifying and analyzing the felt moral responsibility of the ancient physician in regard to his professional role as a potential abortionist or purveyor of euthanasia.

Among the specific questions to be pursued are these. What part did various philosophical attitudes toward death have in shaping the ethical perspectives of Pythagoras, Plato, Aristotle, and Seneca on the above issues? When did these philosophers, in particular, think that human life began and under what conditions, if any, did they allow that human life could be justifiably prevented or ended? Moreover, did members of the Greco-Roman medical and philosophical communities customarily acknowledge an inalienable right to life, or, on the other end of the spectrum, a right to die? Lastly, was suicide admitted to be a rational act in some instances, or was it almost always viewed as impulsive or blameworthy conduct?

Throughout Part Three I shall test my earlier conjecture. I am referring to my claim that many physicians who chose to ignore the teachings of the Hippocratic Oath on abortion and euthanasia (assuming they were familiar with some version of that document at all) did so in part because they were persuaded by powerful ethical arguments which directly opposed the Oath's absolute restrictions against abortion and euthanasia. It is my contention that some philosophical arguments, and the schools from which they issued, undoubtedly did contribute to the pluralistic sentiments of physicians and patients regarding the rightness or wrongness of abortion, euthanasia, and related issues. The task that awaits us is to expose the premises of some of these philosophical arguments and then assess their individual strengths and weaknesses against the backdrop of Greek medicine and popular morals. Moreover, in addition to the preceding quartet of philosophers and their

respective schools, the medical writings of the Hippocratic authors plus
the pioneer obstetrician and gynecologist Soranus shall receive special
attention.[1]

Finally, I must place two qualifications on all that follows. First, I shall
not delve at any great length into the complex subject of ancient law. I shall,
however, comment on legal developments where this is helpful in setting the
stage for my discussion of selected moral perspectives. For it is the moral
perspectives that remain my chief focus here; and so, many details of the law
have been relegated to the accompanying chapter notes.

Second, it will be convenient at times to characterize various ancient moral
positions as possessing consequentialist (teleological) or nonconsequentialist
(deontological) tendencies. Roughly speaking, moral theories are classified
as *consequentialist* if they equate right conduct with acts that cause the
greatest balance of good over evil defined solely in terms of the consequences
that the act in question produces. *Nonconsequentialist* theories, in contrast,
insist that what finally determines the rightness or wrongness of an act has
to do mainly with some other criterion than the consequences which a
given act produces. One example of the former would be ethical hedonism:
pleasure is the highest good, and right acts are those which produce the most
pleasurable consequences for an individual in the long run. An example of
the latter could be called Mosaic ethics: conforming to the divine imperatives
of God is the highest good, and right acts are those which obey the Ten
Commandments — regardless of the specific consequences which these acts
may produce. Strictly speaking, however, such classificatory labels were
foreign to the Greeks; in their moral theory, they generated what moral
theorists today call a uniquely eudaimonistic ethics.

As the words *eu daimon* (good spirit) literally imply, this kind of ethic
involves creating a moral system that aims at making the moral agent a
good spirit — one who acquires well-being. Moreover, contrary to some
criticisms, eudaimonism does not necessarily entail ethical egoism. The
latter states that only those actions which increase a given individual's total
balance of good over evil are morally right and those alone. Yet acts of
charity or benevolence were by no means a standard source of ridicule
amongst the Greeks, for many were aware that the welfare of other citizens
was linked to one's own welfare and to the well-being of the community.
What is common to most all of the Greek moral perspectives about to be
examined, then, is this eudaimonist orientation according to which to do
good (*eu prattein*) is to fare well.[2] Hence the moral agent as a whole, not his
actions alone, are accorded the greatest attention in Greek ethical theories.

In contrast, most modern and contemporary ethical theories have tended (a) to shift the focus of attention from the agent to the praiseworthiness or blameworthiness of his act; and (b) to downplay the notion that the virtuous person is more likely to be the happy person in this earthly existence, at least. These differences in emphasis must be acknowledged at the outset since they are fundamental to the study of Greek moral values.

PRELIMINARY CONSIDERATIONS

Not only abortion, but also infanticide was widely practiced by the Greeks and Romans. In seeking to identify and understand the ancient philosophical arguments for and against these practices, it will be instructive to summarize some of the more interesting and historically important aspects of two practices that were entirely commonplace during most of antiquity. Lecky,[3] Westermarck,[4] and most recently Durant[5] have each written extensive social histories covering Greco-Roman attitudes and practices on these and many related topics from which all may profit. My specific purpose here is to sketch as general background for our upcoming philosophical investigation selected features of the medical, legal, and social landscape that will heighten our sensitivity to Greco-Roman forms of culture and conduct.

The first observation I shall make is that the modern reader who approaches the ancient evidence on these topics for the first time almost certainly faces a radical reorientation of values. One immediately confronts the fact that, with perhaps one known exception, parents throughout Greece and Rome had the legal right (though seldom the obligation) to kill their newly born children for virtually any reason.[6] It was not until the fourth century A.D. that Roman Emperors began to enact measures which sought to protect the lives of free-born infants, at very least.[7] Notice, too, that in a social environment where infanticide is rife and rarely a matter of moral controversy among the general populace, the act of aborting a fetus (so long as the mother is not harmed) tends automatically to be viewed as well within the norm of the morally permissible.[8] In other words, since infanticide was judged morally acceptable under usual circumstances (i.e., when the parents consented, when the infant was still a newborn, etc.), then naturally the decision to carry out an abortion under similarly standard conditions (to be described below) could only with some difficulty be subject to condemnation from the Pre-Christian moral perspective. In a sense, then, the whole subject of abortion for the pagans must be viewed as conceptually dependent on their logically prior reasonings about infanticide. For if infant life was expendable, why protest on moral grounds the killing of fetuses?

What's more, the contemporary reader is at first glance likely to locate the moral question of infanticide as an important branch of the central debate about euthanasia and the prospect of long-term, costly medical care for severely and irremediably defective newborns. But I suggest that a fresh conceptual orientation will need to be temporarily adopted in order to fully penetrate the logic of the pagan mind on this issue. For, as will soon become evident, the reasons most often expressed by the Greeks and Romans for killing their newborn were by no means narrowly limited to modern considerations of what kind of hardship the infant would likely have to endure — given some defect or disease — were he allowed to live. In fact, ancient testimony strongly points to a diverse range of publicly admitted reasons for the justification of either infanticide or abortion. The typical reasons included: (1) the desire to preserve what some women viewed as their sexual, as opposed to their maternal, beauty;[9] (2) the desire to avoid the personal inconvenience and loss of freedom occasioned by caring for very young children;[10] (3) the desire to avoid poverty occasioned by the added material costs of child care;[11] (4) the related desire to limit competition for scarce resources by limiting the size of their community's population;[12] (5) the desire to have male offspring in greater proportion than female offspring based on perceived military and economic advantages;[13] (6) the desire to protect the woman from a life-threatening, difficult pregnancy;[14] (7) the desire to rid society of future citizens who, because they are born weak or defective, are considered to be a worthless burden to themselves or society;[15] (8) the desire to conceal adultery;[16] and (9) the desire to protect one's family or estate from a child of doubtful legitimacy.[17]

It is surely a mistake to imagine that the preceding nine reasons constitute an exhaustive list, or that the average citizens of either Greece or Rome constructed or consulted elaborate philosophical justifications for the above-mentioned deeds. When they acted, they did so partly out of habit and in accord with longstanding cultural approval which they took for granted. Yet it is true, I would argue, that variants of these nine reasons, taken alone or in combination, are highly representative of the sorts of conscious motivations that led individual Greeks and Romans to perform acts of abortion or infanticide.

Nor can one absolutely say, in attempting to come to grips with the approach the ancients took on these matters of life and death, that reasons (5), (7), and (9) alone from the foregoing inventory constituted the usual sorts of justification for infanticide; while (1)–(4), (6), and (8) were standard justifications for abortion. To do so threatens to underestimate the complexity of the ancient, social situation. It overlooks not only the likelihood of unique or mixed motivations for such deeds on the part of the moral agent. But it ignores completely that in a social environment where both abortion *and* infanticide are at very least legal, if not free of controversy, either act may in a given case serve to satisfy the motivation and final aim of the actor.

For example, a married couple may learn that the wife is pregnant. During the pregnancy they may entertain thoughts that they have enough children already and that an additional child will push them further into poverty ((3) above). If they vacillate during the pregnancy on whether or not to abort on account of their poverty, then (3) can still be legally accomplished by means of killing the infant shortly after it is born. There is good reason to believe, incidentally, that infanticide was more likely to result in tentative cases like the one under discussion if the newborn turned out to be a female. For a daughter usually required an expensive dowry to be paid upon her marriage. Also upon marriage, she became subject to the jurisdiction of another household or clan that had contributed nothing to her careful upbringing. Then, too, daughters were expected to do little in the way of outside productive labor that might enrich their family's income (not so, however, in the case of a son). And women in general performed no formal military duties in defense of their family or their community.[18] Hence there is little reason to doubt that throughout antiquity many more female than male children perished by infanticide. In Hellenistic Greece, the average family was quite small. It rarely contained more than one daughter. Thus it is fair to say that the comparative lack of value placed on female infants can be plausibly understood as the predictable outcome of an uncompromising and dominantly utilitarian appraisal of their worth.[19]

The preceding account of the social and legal setting in Greece and Rome has to this point been characterized in the broadest fashion. Again, my purpose is to provide but a general picture of relevant Greek and Roman popular thought and customs in preparation for introducing the philosophical literature which it is the main business of Part Three to explore. Yet there remain three additional areas bearing on our topic which deserve some comment. These areas are: (1) contraception in antiquity; (2) methods of abortion and infanticide; and (3) the medical role of midwives.

Contraception

By the fifth century B.C., if not long before, the Greeks were generally aware of the causal connection between sexual intercourse and human pregnancy. In fact, embryology was a topic to which most of the sixth and fifth century Pre-Socratic philosophers had directed at least a passing comment.[20] Yet it was not until the time of Aristotle, and the Hippocratic authors of *On Semen* and *On the Development of the Child*, that human embryology became a distinct biological science wedded to the search for nature's impersonal laws. Furthermore, though it is difficult to estimate how evenly distributed this embryological knowledge was among the general populace, it is certain that various sorts of contraceptive measures were known and used by the citizens of Greece and Rome.

These measures, however, were usually not reliable. If they worked at all, very often even their authors did not correctly understand why. Thus Norman Himes, in his *Medical History of Contraception*, states that while Aristotle is the first Greek on record to mention a contraceptive formula — a formula which involves the application of oil of cedar to the womb prior to intercourse — there is little reason to believe that he fully understood the causal principle behind this supposed remedy. "It is clear from the context that he regarded the quality of smoothness as the *modus operandi* of prevention; whereas we now know that oil has a contraceptive effect by reducing the motility of spermatazoa and by gumming up the external os."[21] In any case, Himes grants that Aristotle's formula might have worked on occasion. In addition, there were the more widely used mechanical devices such as wool, salves, and resins. These, in the opinion of the physician and medical historian Henry Sigerist, were probably somewhat more effective than the potion Aristotle mentions above when introduced into the vagina before intercourse.[22]

What would have constituted the ideal contraceptive device for both ancient and contemporary times, had it worked, was erroneously prescribed as follows by the Hippocratic author of *On the Nature of Women*:

If a woman does not wish to become pregnant, dissolve in water misy [a copper salt] as large as a bean and give it to her to drink, and for a year she will not become pregnant.[23]

Furthermore, the Roman poet and Epicurean philosopher Lucretius (99—55 B.C.) declares at the end of the fourth book of his widely admired *On the Nature of Things* that the gods are definitely not responsible for human infertility and sterility. Instead, he says, natural causes are to blame. Hence he mocks desperate childless couples for appealing to the gods, and scolds husbands in particular for " . . . piling the raised altars with offerings, to make their wives pregnant with abundant seed. In vain they weary the divinity of the gods and the sacred lots." [24] Lucretius identifies the causes of infertility as including "too great thickness" or "undue fluidity and thinness" of the semen.[25] He also suggests that both diet and the modes of intercourse can improve or diminish one's chances of conception. For the consistency of semen is, he thinks, affected by certain foods; in addition, the quadruped position is said by Lucretius to promote conception. As for methods of contraception, he wrongly states that "effeminate motions" or gyrations on the part of women during intercourse help prevent conception. Lucretius thus concludes that " . . . for their own ends harlots are wont to move, in order not to conceive and lie in child-bed frequently, and at the same time to render Venus alone attractive to men. This method our wives have surely no need of [if conception is desired] ."[26]

Lastly, the most accurate and impressive account of contraception in antiquity is rendered by the Greek physician of the methodist school, Soranus (A.D. 98–138). Soranus is universally credited as an early pioneer and founder of the medical specialties of gynecology and obstetrics. More clearly than his contemporaries, he distinguished between contraceptives and abortifacients, saying, in part:

A contraceptive differs from an abortive, for the first does not let conception take place, while the latter destroys what has been conceived. Let us therefore call the one 'abortive' (*phthorion*) and the other 'contraceptive' (*atokion*).[27]

Also, far in advance of his time, Soranus warned about the possible health risks involved in *drinking* alleged contraceptive potions of the sort cited by the Hippocratic authors above. He resisted superstition at every turn, and discouraged the use of amulets (sometimes made in the form of genital organs) which were thought by the masses to influence the prevention or promotion of conception. According to Himes, Soranus also prescribed a host of ineffective contraceptive formulas among the 150 or so such prescriptions that he gave. But his main reliance was on the more rational mechanical techniques such as vaginal plugs or pessaries, which used wool as a base mixed with oils or other gummy substances. As for his related views on the morality of abortion, which will be examined in the discussion below, Soranus observed: "It is much more advantageous not to conceive than to destroy the embryo. . ."[28]

Given the results of our brief examination up to this point, I suspect that most Greeks and Romans would have shared Soranus' stated preference "not to conceive" rather than resort to abortion or infanticide. The Greeks and Romans were not a bloodthirsty people, and they were well aware of the risks and inconveniences attendant upon either remaining option.[29] But even so, universal ignorance of truly effective contraceptive measures would have rendered any such moral preferences entirely moot. As has been suggested, ancient medical science neither knew nor communicated any really reliable contraceptive methods. It is generally agreed that the most effective and widely used method available was *coitus interruptus*. In addition, I am persuaded that the most our historical evidence entitles us to say is that the decidedly flawed contraceptive knowledge of antiquity was known to the more privileged segments of society, to be sure. But how widely dispersed these detailed formulae were among the uneducated and highly superstitious majority is hard to gauge. It is entirely possible that the masses experimented with an endlessly changing list of equally ineffective contraceptive techniques of every imaginable type and origin.

But, if so, we must confront an additional important fact: limiting family size by whatever means was a widespread ancient practice. We possess several reliable reports, especially during the Hellenistic and Roman periods, of whole communities facing much lower population counts than their leaders and their own social critics deemed desirable from the standpoint of the national welfare. In Macedonian times, for example, Philip V was so worried about his country's depopulation that he created a law which forbade the willful limitation of the family by contraception, abortion, or infanticide. It is reported that in thirty years Philip's edict succeeded in raising the manpower of his nation by almost fifty per cent.[30] Polybius, writing about 150 B.C., was well aware of the widespread practice of family limitation. He complained bitterly and wrote in alarming tones about the resulting depopulation:

In our time the whole of Greece has been subject to a low birth rate and a general decrease of the population, owing to which cities have become deserted and the land has ceased to yield fruit . . . For as men had fallen into such a state of luxury, avarice, and indolence that they did not wish to marry, or, if they married, to rear the children born to them, or at most but one or two of them, so as to leave these in affluence and bring them up to waste their substance – the evil insensibly but rapidly grew. For in cases where, of one or two children, the one was carried off by war and the other by sickness, it was evident that the houses must have been left empty . . . and by small degrees cities became resourceless and empty.[31]

Can one fairly surmise from such testimony that the most common methods of voluntary family limitation in antiquity were abortion and infanticide, and not the relatively ineffective contraceptive measures of the sort outlined above? In my opinion, this is the only acceptable conclusion. Of the well-to-do families, Poseidippis (fl. 270 B.C.) writes: "Even a rich man always exposes a daughter."[32] Famine, sickness, poverty, and war, however great their toll, could not alone account for the reported facts. And, as we shall soon see, several Greek philosophers of the period condoned abortion or infanticide as morally justifiable practices for the control of the untoward social consequences of overpopulation. Is it not reasonable to think, therefore, that the public endorsements of philosophers may have influenced at least some segments of Greco-Roman society?

Methods of Abortion and Infanticide

Our investigation ought not to advance any further until we pause to define our two principal terms in this section. For our purposes, abortion shall be defined as the deliberate termination of pregnancy resulting in the intended death of the fetus prior to normal or spontaneous delivery.[33] Probably the most effective method, if not most popular, was to puncture the ovum by manual or other mechanical manipulations.[34] Another method, recommended by a Hippocratic author to a singer who unintentionally became pregnant, was to have the woman jump high into the air, kicking her heels repeatedly into her buttocks.[35] Embryotomy, the surgical cutting apart of the fetus, was also known to both the Greeks and Romans. But this was apparently a procedure of last resort. It was used by physicians primarily when the mother's life was threatened during delivery or when the fetus was thought to be dead *in utero*. Additionally, there were numerous recipes for abortifacient drugs taken orally or as pessaries (vaginal suppositories) which were touted by various medical writers. But most of these have since been shown by modern medical analysis to have had little effect.[36]

In contrast, the question of how to define infanticide an a way that conforms to ancient Greek and Roman thinking initially poses more of a problem. Straightforward infanticide, then as now, is commonly taken to mean the willful termination of the life of an infant. This definition may serve to capture standard cases in antiquity where the infant was *actively* killed by such common methods as smothering, strangling, drowning, crushing, stabbing. But this definition fails to capture the subtleties involved in the allied and more passive methods vaguely called exposure or exposition.

Why? Because exposure sometimes involved merely acting to rid oneself of the burdens and liabilities of the infant, short of actively killing it. Hence to say in antiquity that an infant had been exposed by its parents was not always the same as saying that the parents killed, or even intended to kill, their infant.

For example, a very common method of exposure in Greece and Rome was to carefully place the infant in an earthenware vessel and abandon it near a religious temple. The exposed infant was sometimes saved by some benevolent person wishing to adopt, or by someone hoping to profit by raising the child to serve as a slave or a prostitute.[37] Reflection on the latter possibility apparently inclined some parents to actively kill their unwanted infants rather than tempt fate to condemn such children to a wretched life of poverty or bondage at the hands of others. Thus, in one of the few characterizations of infanticide in Greco-Roman imaginative literature, the Roman playwright Terence (c. 190–159? B.C.) focuses on just such a concern. He portrays a husband and wife who are torn over what to do with their unwanted infant.[38]

Terence represents Chermes as having, as a matter of course, charged his pregnant wife to have her child killed provided it was a girl. The mother, overcome by pity, shrank from doing so, and secretly gave it to an old woman *to expose it, in hopes that it might be preserved*. Chermes, on hearing what had been done, reproached his wife for her womanly pity and told her she had been not only disobedient but irrational, for she was only consigning her daughter to the life of a prostitute.[39]

The moral and psychological significance of this distinction between infanticide, taken here to mean the active and certain termination of infant life, and exposure, taken here to mean passively subjecting the infant to the increased likelihood — but not always the certainty — of termination, cannot be overlooked. Psychologically, the uncertainty of the outcome of exposure in standard cases of abandonment, such as those described above, probably relieved some parents of their sense of moral responsibility for their deed. After all, whatever the outcome, it could be thought of as ultimately a matter now in the laps of the gods. Heaven could decide. Morally, it is quite possible that exposure was not popularly considered to be as blameworthy an act as outright infanticide. This was particularly so in regard to the infanticide of *healthy* infants, which custom, if not law, generally opposed. Thus, by the period of the Roman Empire, at least, the chances of healthy infants being rescued from abandonment were somewhat improved owing to the occasional intervention by the state.[40]

Midwives

No inquiry into Greco-Roman values and customs associated with the birth of new human life would be complete without some mention of the medical and social roles of midwives. It is probably true that most Greek and Roman women gave birth in a squatting position, perhaps sitting on a specially designed parturition chair. If medical services were rendered at all (and in most cases they were not thought to be needed), it was typically the midwife who attended. Socrates, himself the son of the midwife Phaenarete, is made to say at *Theaetetus* 149D that among the things midwives are known to do well is " . . . at an early stage to cause a miscarriage if they so decide."[41] Sigerist credits them with being able to diagnose early pregnancy better than any physician, and he speculates that midwives called in physicians during the birth process " . . . only in abnormal cases when something went wrong."[42]

It is not known with precision what, if any, the role of midwives may have been in association with acts of infanticide or exposure. But in the matter of procuring abortions, we have sufficient evidence to suggest that a good many midwives helped to provide these services for a fee. And this they did not only in cases of possible therapeutic abortion, i.e., those done to protect the mother's health — like the type which Socrates may have been referring to above. On this we also have the testimony of Soranus, who considered the subject of midwifery to be deserving of special comment. "He emphatically demands of the ideal midwife that 'she must not be greedy for money, lest she give an abortive wickedly for payment,' a warning that would hardly have been necessary unless the temptation [to perform abortions] had been great in ancient Rome."[43]

Hence the moral implication of the role of midwives in ancient family life was this. Since midwives could detect pregnancy in the early stages, and could provide a woman with a relatively safe abortion, it was in principle possible for a woman, either married or unmarried, to secure an abortion without the knowledge or consent of her lover, her family, or her community. That this, in fact, sometimes occurred is well documented by certain Hellenistic legal prohibitions which presuppose such behavior.[44] A second moral implication here bears on the Hippocratic Oath. When the Oath-taker states, "I will not give a woman an abortive remedy," he is at once tacitly disassociating himself from the practices of many midwives with whom he does not agree, based on moral considerations predicated on at least some Pythagorean perspectives.[45] To these Pythagorean and other philosophical perspectives on abortion we now turn.

PHILOSOPHICAL PERSPECTIVES

The principal philosophers or schools of philosophy on which I have gathered
ample evidence to permit a reliable identification and subsequent discussion
of their respective stands on abortion or infanticide are just four in number:
Pythagoreanism, Plato, Aristotle, and the Stoicism of Seneca. Whenever
possible, I shall try to determine in each case the answers to two basic questions. First, at what point is the developing fetus taken to be a human being?
Second, on what moral grounds is abortion approved or condemned? I shall
limit the first question primarily to the metaphysical or early scientific
considerations of when the preceding four authors judged the human embryo
to be ensouled or animate. This was, after all, the characteristic form that
that question took among the ancients. In addition, I shall limit the second
question primarily to the judgments of value placed on fetal or infant life.
While some attention will eventually be paid to the more global question of
what value was placed on adult human life *per se*, that is a broader issue
which will be postponed until the remaining question of euthanasia is taken
up in the next chapter.

Pythagoras

It will be recalled that Pythagoras (c. 580–497 B.C.) placed the highest value
on the soul. The earthly soul is said to be a temporarily fallen divinity, immortal in character, and the most essential and enduring part of each person's
identity. In our discussion of Greek attitudes toward death in Chapter Three,
we saw that the Pythagoreans believed in what I there called *divine personal
immortality*. We also saw that the soul is yoked to the alien and inferior
human body (*soma*) for the sake of punishment, as in a tomb (*sema*). The
personal salvation promised to humankind by the Pythagorean Brotherhood
was that for the properly initiated at least, living on earth in accordance with
the Pythagorean ethical virtues could render one's soul sufficiently pure in
character to assure its speedy return at death to the heavenly station from
which it originated. Again, the body alone is subject to decay: the soul, being
divine, is deathless and is the object of highest worth from the Pythagorean
perspective.

   To these metaphysical and religious tenets the Pythagoreans added the
dogma of the transmigration of souls. This stated that upon death, if one's
soul was not sufficiently pure to return to its natural place among the gods in
the heavens, then it would be required to inhabit fresh (newly born) bodies,

of either human or animal form, till such time as the requisite moral purity had been achieved. Thus, given the high value placed on the soul, it is little wonder that the Pythagoreans held it to be morally wrong to kill human or animal life. For while it is true that they downgraded the body and held the soul to be superior, they also held that it was divine law that decreed what kind of body, if any, a soul must occupy.

As we shall see in our upcoming discussion of euthanasia, it was further held by them to be one's duty neither personally to take early leave from one's body (say, by suicide), nor to assist or coerce any other living person or animal in doing so. To do so was both to mock the gods and soil one's soul with evil. This religious reasoning justifies at once, therefore, the Pythagorean observance of certain blood taboos such as the abstention from eating meat. More importantly, it also justifies the judgment that it is morally wrong to cause a woman to have an abortion *if, in addition*, fetuses are taken to possess souls.

What, then, was the Pythagorean teaching on the question of the status of the fetus? As suggested in our preliminary discussion of abortion in connection with the Hippocratic Oath in Chapter Five, the Pythagorean belief was that the fetus possessed a soul and was an animate being from the very moment of conception. Though the evidence for this claim cannot be traced directly to Pythagoras – since we possess nothing that he wrote – at least two pieces of evidence seem to justify this conclusion. The first is the testimony of Porphyry (A.D. 232/3–c. 305), who, in his treatise on embryology, expressly attributes this position to the Pythagoreans.[46] The second is based on a Hellenistic account of the Pythagorean system of physiology compiled by Alexander Polyhistor (c. 105 B.C). On the latter account, the germ or seed is said to be a piece of brain containing hot vapors within it, and soul and sensation originate from this vapor at conception.[47] Similar views were also previously accepted by the Pythagorean Philolaus in the fourth century B.C., as reported by Diogenes Laertius.[48]

Thus I see little reason to doubt the claim that the Pythagoreans unconditionally condemned abortion. As far as our evidence implies, fetal life was co-equal in moral worth with adult human life, and this from the moment of conception. If so, this finding would appear at once to rule out the permissibility of therapeutic abortions for the Pythagoreans. For if the soul of both the fetus and the pregnant mother is divine, and if it is a decree laid by the gods upon mankind that the foreordained schedule of the arrival and departure of souls must never be tampered with by human beings, who, without contradiction, can claim that killing the fetus to save the mother's life is justified? Moreover, who would claim, granting the identical premises, that terminating the life of infants — for whatever reason — is on Pythagorean principles justified?

In addition, it is also worth noting that, on the above interpretation, not only human abortion and infanticide but any effort to terminate the embryonic or newborn life of any species of animal would be subject to moral condemnation by the Pythagoreans. As Xenophanes, a contemporary of Pythagoras, mockingly recounts:

They say that once a *puppy* was being whipped, Pythagoras, who was passing by, took pity on it, saying: 'Stop! Do not beat it! It is the soul of a friend; I recognize his voice.'[49]

This concern for the lives of human as well as nonhuman creatures further illustrates a point that should not be skated over. That is, in talking about the concept of respect for life (also sometimes called the concept of sanctity or reverence for life), it is useful to distinguish between a *comprehensive* respect for life and a *noncomprehensive* respect for life.[50] The former covers all forms of life, vegetable, animal, human. The latter is more representative of Western thinking: it is restricted to respect for life in the human form alone (whatever that designation is taken to be). In the case of the Pythagoreans, however, it is fair to say that their strong respect for life ethic approximated a more comprehensive scale than did any of the competing philosophical schools throughout antiquity.[51] Therefore, this ethic constituted a distinctive trait of Pythagorean moral and religious thinking — one that was sure to inform their equally nonconsequentialist stand against suicide and euthanasia.

Plato

The views of Plato on abortion and infanticide, like those of Aristotle, are predicated on considerations of a far more utilitarian and this-worldly sort than those associated with the Pythagoreans. (I am here using the word "utilitarian" in the broadest sense to mean the social consequences of an act or policy; this sense is not to be confused with the much later utilitarian school of philosophy associated with Bentham and Mill.) In a way, this is at first glance a bit puzzling in the case of Plato. As we have already seen, Plato believed in a version of *divine personal immortality*.[52] He also accepted the dogmas of the soul-body dualism and the transmigration of souls that had been associated with Orphic-Pythagorean thought. How, then, could he

Plato → life begins @ birth

endorse both abortion and infanticide in his vision of the ideal state as set forth in his *Republic*?

The answer to a large extent lies in the fact that, unlike the Pythagoreans, Plato held that human life began at the moment of birth. This, at least, is the contention of Edelstein. While I have been unable to discover an explicit statement to this effect in Plato's own dialogues, his mythical account of the soul's journey into successive human lives, known as the Myth of Er, does seem to imply this interpretation.[53] Moreover, that the successors of Plato at the Academy took this position is almost certain. Platonists and Neo-Platonists alike denied the ensoulment of the fetus *in utero*. This they did on the grounds that the soul, being the special divine entity it is, had to enter the body from without (perhaps through the newborn's first breaths). Given this embryological theory, it is then possible to assert that aborting fetal life is an act falling in an altogether different category than killing a human being. For the fetus remains a soulless creature.

What, then, are Plato's reasons for justifying *abortion* as he sets forth his blueprint for utopia in the *Republic*? In his organization of marriage and family life among his Guardian and Craftsmen classes, he accords the highest value to what he deems the good of the state. Ultimately, that highest value is identified as justice, the crowning virtue which emerges in the whole community when subordinate virtues are achieved in the right proportion among all the citizenry. If the state is to be the best, then its future citizens must not be allowed to live unless they can serve the productive interests of the state at some minimal level of competence. In the main, it is on this policy that Plato attempts to justify abortion, infanticide, and euthanasia. For he does so, essentially, by tacitly appealing to the consequentialist principle that to permit these three acts in certain circumstances promotes the greatest long-term benefit for the entire community. Thus, in the final analysis, whatever rights or duties the individual may have to himself or to his family, these are subordinated by Plato to the higher duty to serve the state.

Specifically, Plato authorizes abortion as part of his eugenic program. This program entails the strictest regulation of marriage and mating, with the end in view of producing the most perfect human offspring possible for the alleged benefit of the entire state (*Rep.* V, 459). Wishing to maximize the quality of the offspring of the Guardians in particular, he specifies that Guardians can legally mate only between the ages of 25 to 55, and 20 to 40, for male and female partners respectively. He prescribes these age ranges since he thinks they conform to the eugenic principle " . . . that children should

be born from parents in the prime of life."[54] Then, in regard to abortion,
Socrates states:

As soon, however, as the men and the women have passed the age prescribed for producing children, we shall leave them free to form a connection with whom they will . . . ;
and all this only after we have exhorted them to see that no child, if any be conceived,
shall be brought to light, or, if they cannot prevent its birth, to dispose of it on the
understanding that no such child can be reared.[55]

It should also be observed that Plato's eugenic policy, when extended to
cover the special case of fetuses conceived out of wedlock by women under
the prescribed age of 20, almost certainly entails the termination of these
illicit pregnancies as well. For Plato held that offspring conceived by parents
who were either too young or too old faced an unacceptably high risk of
possessing inferior physical or mental attributes. This end, i.e., attaining the
best bred and best educated Guardians possible, he thought morally justified
these means.

So, too, Plato strongly implies that any fetuses conceived by lustful citizens
who fall within his approved age ranges, but who, in addition, fail to secure
the required legal and religious sanctions of the state (marriage), must thereby
suffer abortion. This, even against the parent's will. Hence all instances of
bigamy, adultery, incest, and pre-marital sexual intercourse that produce offspring will be subject to abortion or infanticide.[56] In other words, illegitimate
offspring were not to see the light of day. This suggests that Plato's legal
sanctions for sexual incontinence or disobedience were grounded on eugenic
principles, as well as on independent calculations of the overall good to be
achieved by the anticipated deterrent value of these strict sanctions. For it
is not hard to imagine that, in some of these illegal unions, offspring of
a potentially high quality from the strictly genetic standpoint would result.
Nonetheless, they are put to death by the state as a public warning to other
like-minded citizens who would flirt with such folly.[57]

That Plato also favored *infanticide* for the purpose of securing the best
possible Guardian class, and that he specifically recommends getting rid of
defective newborns judged to be below some minimal standard of fitness
(which he does not specify in any detail), there can be little doubt. In a key
passage he reveals that:

As soon as children are born, they will be taken in charge by officers appointed for the
purpose, who may be men or women or both, since offices are to be shared by both
sexes. The children of the better parents they will carry to the crèche to be reared
in the care of nurses living apart in a certain quarter of the city. Those of inferior parents

and any children of the rest that are born defective will be hidden away, in some appropriate manner that must be kept secret.[58]
They must be, if the breed of our Guardians is to be kept pure.

It is more difficult to determine, however, what additional reasons, if any, Plato supplied for the taking of infant life over and above the utilitarian reasons he advances for the taking of fetal life. On the heels of the passage just quoted, he supplies no new grounds. And I know of no other text where he does so in specific reference to infanticide, even though we might well expect additional reasons since he accords to the infant a higher moral status than he does to the fetus. But there are some hints of a possible reply, I think, in remarks that he makes at *Republic* 406 concerning the importance of living a productive life and the moral justification he advances there for euthanasia among the incurably sick or severely disabled adult population. Since it will be more convenient to explore this last matter a bit later on, it will suffice here to remark that there is an apparent weakness in Plato's ready endorsement of infanticide. This weakness stems from his inattention to taking into full account the possible divergent moral status of infants as compared with fetuses.

Aristotle

Arguing along utilitarian lines similar to Plato, but emphasizing to an even greater extent the threat of overpopulation to the well-being of the state, Aristotle judges both abortion and infanticide to be morally justifiable. According to Aristotle, "if no restriction is imposed on the rate of reproduction, ... poverty is the inevitable result; and poverty produces, in its turn, civic dissension and wrong-doing."[59] Aristotle, in other words, tacitly argues that since there is a causal link between overpopulation and occurrences of poverty or political revolution, and since these are social consequences that seriously threaten to reduce the overall balance of good and stability in the state, then the imposition of limits on family size is morally defensible. Such limits, he insists, must in any case be adjusted to allow for (1) infant mortality rates and (2) normally expected infertility rates among the married.[60] Hence it is fair to conclude further that, given the above reasoning, Aristotle would not have objected to couples' resorting to contraceptive measures in order to achieve ideal population levels within the state. In fact, as we saw in a preceding section, Aristotle was well aware of such devices, and voiced absolutely no objection to their use.[61]

In this connection, his position on abortion and infanticide is a bit more complicated than the comparatively unqualified approval given it by Plato in the latter's morally eligible cases. In Book VII, 1335b 20, of his *Politics*, Aristotle states:

> The question arises whether children should always be reared or may sometimes be exposed to die. There should certainly be a law to prevent the rearing of deformed children. On the other hand, there should also be a law, in all states where the system of social habits is opposed to unrestricted increase, to prevent the exposure of children to death *merely* in order to keep the population down. The proper thing to do is to limit the size of each family, and if children are then conceived in excess of the limit so fixed, to have miscarriage induced *before sense and life have begun in the embryo.*[62]

First of all, in the prior case of infanticide, Aristotle deems this justifiable when and only when an infant is born deformed. The kind and degree of deformity he has in mind is nowhere specified. Thus it would be misleading, I think, to attempt to associate Aristotle with a specific and perhaps overdetermined criterion on this matter. It must suffice to observe that he often stresses the importance of a sound bodily and mental constitution in his writings, as I suggested in a previous chapter. Also, he counts as life's chief non-moral goods (i.e., qualities which contribute significantly to living the good life) both good birth (*eugeneia*), which includes sound heredity; and health (*hygeia*), which includes freedom from disease and the full possession of bodily faculties.[63] Given that a deformed infant might well be judged to have little purchase on either or both of these valued attributes, one again observes Aristotle's concern for the supposed greater good of the whole community in his approval of infanticide. For a chronically sick, weak, or deformed child can, on utilitarian grounds, be judged of little worth either to himself or to the long-term productive interests of the well-ordered state.

But just what Aristotle's position is on abortion is somewhat harder to interpret. In view of the preceding quotation, his position appears to depend almost entirely on when, according to his embryological and related metaphysical speculations, the human zygote was considered by him to have acquired "sense and life." In order to answer this question fully, it will be necessary to supplement what we have already learned concerning Aristotle's views on the human soul during our prior discussion of his concept of death.[64] We noted there that Aristotle regarded the soul to be inseparable from the body. He also taught that all living things, *zoa*, are single concrete entities comprised of soul and body. The soul he equated with the form of a given species. (Literally, soul (*psyche*) in Greek meant "life" or "vital principle;" it was then a word less closely associated with the religious connotations it

Abortions acc. to Aristotle are rt. only if a family has exceeded the # of kids they can have in order for society not to be overpopulated

tends to have now in English).[65] In addition, Aristotle taught that nonliving or inanimate things are soulless (*apsychon*). Hence the possession of a soul made all the difference in Aristotle's system in regard to whether an entity of a given kind was to be sorted into the living (literally besouled (*emsychos*)) or nonliving categories.[66] Observing this, David Ross has written:

> Aristotle is helped here by the fact that the natural expression in Greek for a living thing is *empsychon soma*, 'besouled body,' where 'besouled' evidently stands for the attribute that distinguishes living from other bodies, the power (at the least) of self-nourishment with or without the powers that mark off the higher living things from the lower. Soul, then, is the form or actuality of a living thing [whereas body is its potentiality].[67]

In reference to Aristotle's view on the status of the human zygote in his theory of the ascending scale of being, it is further necessary to ask: What, in the broadest terms, distinguishes higher living things from lower? Briefly put, his answer is that all living things are possessed of souls, and of souls there are three basic kinds. These are, in ascending order of worth: (1) the nutritive soul; (2) the sensitive soul; and (3) the rational soul. These three forms of soul constitute a series with a definite order, such that each type presupposes all that come before it in this order, without being strictly implied by them.

Hence the minimal soul ascribable to all living things is the nutritive soul. The human zygote, or conceptus, is on Aristotle's view at very least besouled in the nutritive sense, for it is capable of nourishment and growth. So, too, all plant life is in this sense "alive." But plants, unlike persons, are limited to this nutritive level of existence.

Higher up the scale of being are creatures to whom the sensitive soul is properly ascribable. All animals are, at very least, besouled in virtue of their capacity to touch, and so they feel pleasure and pain, and also desire.[68]

Finally, there is the rational soul which is ascribable to man alone, of all earthly creatures. I conclude that to be a human being in the fullest sense means for Aristotle to possess the capacity for reason, contemplation, and rational choice. If there is to be talk about degrees of *human* existence, for Aristotle, such talk must ultimately be referred to various stages of growth, development, and eventual decay on the preceding account of his psychology.

In regard to his theories of human embryology and psychology, Aristotle's position on the morality or immorality of abortion may now be properly sorted out. It will be recalled that he condemns infanticide " . . . *merely* in order to keep the population down." He then says that the proper or right thing to do is to secure an abortion only if a given family has exceeded its fixed quota of children. It is significant, I think, that he does not cite any other immediate reasons than population control as grounds for abortion. Thus such possible grounds, common enough in his time, as the desire to preserve feminine beauty, avoid personal inconvenience, conceal adultery, etc., he does not even mention. Hence I am inclined to think that he considered these motives to be morally ineligible.

At any rate, though he justifies abortion in the lone case mentioned, he does so by adding a strict condition which, if it is not met, renders the deed morally objectionable. That condition is, in Aristotle's own words " . . . to have miscarriage induced before sense (*aisthesis*) and life (*to zen*) have begun in the embryo." We are now in a position to see that by life, Aristotle was referring minimally to a living thing possessing the nutritive soul. This, as I have suggested, is the status of the human zygote at and shortly after conception.[69] Sometime later — Aristotle nowhere says how long but perhaps in a month or so — the developing fetus is informed by the sensitive soul ascribable to all animal life. Aristotle remains curiously silent as to the precise moment *in utero* when the fetus acquires the rational soul, though this was usually conjectured to occur a bit later on during the second trimester of pregnancy.

Let us suppose, as did his followers in antiquity, that some point between the fortieth and ninetieth day of gestation marked the cut-off date during which induced abortion (for the lone eligible reason of family limitation) became morally wrong. This, supposedly because abortion then involved the destruction of a fetus endowed with the animal capacity to feel pain. In other words, abortion was then wrong because it involved the destruction of sentient *and* potential human life. If we suppose this, we come at once to a dilemma. The dilemma is that in his *History of Animals*, Aristotle states:

In the case of male children the first movement usually occurs on the right-hand side of the womb and about the fortieth day, but if the child be a female then on the left hand side and about the ninetieth day. However, we must by no means assume this to be an accurate statement of fact, for there are many exceptions . . . . And in short, these and all such-like phenomena are usually subject to differences that may be summed up as differences of degree.
About this same period [40 days for the male, 90 days for the female] the embryo begins to resolve into distinct parts, it having hitherto consisted of a fleshlike substance without distinction of parts.[70]

Now if one is willing to assume, as did some of his followers, that Aristotle is here implying that the fetus acquires the sensitive soul at forty and at ninety days in the case of the male and female embryo *respectively*, then how did the moral agent in the relevant case proceed? For it was impossible in antiquity to know in advance the sex of the unborn with any precision, as Aristotle here admits himself.

There is the additional problem, which Aristotle also acknowledges, that not all women can know within forty days that they are truly pregnant. If so, is it always too late to justifiably abort after forty days (waiving here the added difficulty of ascertaining the exact date of conception)? It would appear so. For, again, if one waited until, say, the sixtieth day after suspected conception, and if to one's surprise the fetus was identifiable as male, then on Aristotle's view one has acted improperly. Therefore, it is safe to infer that the most conscientious thing a married couple ought to do, according to Aristotle, is to adopt birth control methods of some sort in cases where the family quota of offspring has already been reached. If these devices failed, as they commonly did in his day, then abortion in the eligible case becomes morally acceptable only if both of the following conditions are met: (1) the date of conception can be reasonably determined; (2) the woman realizes that she is pregnant and aborts *prior to the fortieth day* of gestation. From this I conclude that Aristotle's moral position of the question of abortion is fairly clear: he gives it limited approval on utilitarian grounds. But practical difficulties of an epistemological nature concerning the dilemma of accurately pinpointing the date of acquisition by the human embryo of the relevant sensitive soul effectively casts Aristotle's moral stand on abortion into a sea of confusion. For if one assumes, for the sake of argument, that it is in fact morally justifiable to abort during some favored period of time t_1, and not at successive time t_2, but there is no reliable way to verify whether one is aborting within the favored time period, then practical problems in ethical decision-making are bound to result.

Therefore, as to Aristotle's views on abortion, I see no natural way out of this dilemma. Indeed, it is probable that he himself was aware of these epistemological difficulties. But even though he made an attempt, he apparently found no fully satisfactory solution to them.[71]

Seneca

The Roman Stoic, Seneca (94 B.C.–65 A.D.), conveniently aids our initial approach to the Stoic perspective on abortion and also serves to anticipate his own position on euthanasia (to be discussed later on). In one telling passage, he states:

Just as the mother's womb holds us for ten months, not in preparation for itself, but for the region to which we seem to be discharged *when we are capable of drawing breath* and surviving in the open, so in the span extending from infancy to old age, we are ripening for another birth . . . The day which you dread as the end is your birth into eternity.[72]

In actuality, Seneca felt that the best state of affairs was never to have been born in the first place.[73] This did not mean that he had no regard at all for human life. Rather, he shared a sentiment common to many Stoics and Epicureans who lived during the Hellenistic and Roman periods. This somewhat nihilistic sentiment stemmed in large measure from the growing feeling that the satisfactions of human existence are ordinarily too short and too few. Life holds many apparent pleasures, to be sure, but they seldom abide owing to their frail and transitory nature. Even those things that, on the Stoic view, men and women ought to prize most of all are extremely hard-won and always subject to loss: peace of mind (*apatheia*) above all else; personal and communal well-being; a sense of at-oneness with God and Nature; and, to a lesser extent, personal health.

The means to spiritual peace of mind, which the wise man or woman alone could hope to sustain but which all might experience to some degree if they lived life in the right manner, was constituted by the four cardinal virtues. These four virtues were identified by the Stoics as: (1) wisdom (knowing what is truly good and bad); (2) courage (knowing what to truly fear and what not to fear); (3) justice (knowing how to give everyone what properly belongs to him); and (4) self-control (knowing what impressions to assent to and what passions to moderate or ignore). These virtues, and their supposed benefits, were also quite representative of the eudaimonist ethical perspective according to which to live virtuously promotes the well-being of the whole person — both his body and soul, his commerce and relations with others, and his sense of unity with the greater universe of which he is but a small part.

Returning to Seneca's preceding quotation, it is evident from his words that he associates being born with the initial capacity of the individual to draw breath. This turns out to be an important observation. For, according to later Stoic physics and psychology, (a) the soul is a corporeal thing; and (b) its essence, and indeed the defining attribute of being a member of the human species, is the possession of the vital breath (*pneuma*).[74] As pantheists, the central Stoic teaching in this connection was that each person's soul is a tiny, fiery piece of the greater world-soul which is God. Hence, on the Stoic perspective, birth marks the start of human life in virtue of the manifest capacity of the newborn to draw breath.[75] (The fetus is ontologically akin to a mere plant: it possesses a *physis*, a nature, but not yet a *psyche*, a soul.)

Breath or *pneuma* was also said to be fed by exhalations from the blood. It was taken to be diffused throughout the entire unified soul (though the heart was its principal seat) owing to the interpenetration of breath-currents throughout the body. Human death, therefore, is signaled by the absence of breathing. It is occasioned by the loss of *pneuma* which, on the Stoic view, was causally linked to the final cessation of cardiopulmonary functions. The dominant Stoic teaching about the dead, then, involved a variant of *natural personal dissolution*. They had it that the deceased's *pneuma* reunites in death with the larger *pneuma* of the world-soul, extinguishing thereby any trace of personal individuality.

Given this orientation, it comes as little surprise to learn that generally the Stoics were not adamantly opposed to abortion. After all, the fetus was not counted as a human being in either the biological or moral sense. Here a useful contemporary distinction may be introduced. One may talk about humanhood (or personhood) in two senses: (1) in the *biological* sense, designating the essential properties a being must have to count as a member of the species *homo sapiens*; and (2) in the *moral* sense, designating the properties a being must have in order to be counted a member of the moral community entitled to certain rights, duties, and related legal or moral protections such as, for example, the rule against homicide.[76] Roughly, many of the Stoics of Seneca's period believed that the human soul, following birth, continued to expand its powers until around the age of 14 years it acquired the full capacity of rationality.

Fundamentally, the Stoics held that justice only extends to rational beings. In addition, only rational beings could be the bearers of rights. Therefore, only those 14 years or older were considered mature enough to possess rights protecting them against homicide (among other things). So while personhood in the biological sense was acquired at birth, personhood in the moral sense was acquired only at the point of young adulthood in the normal individual. The fetus, though not totally worthless, is entitled to our moral concern mainly insofar as nature intends fetuses to become rational beings in the usual scheme of things. Thus, since one ought to follow in general the designs of nature, it would be wrong to adopt a frivolous attitude toward the act of abortion.

However, even though children are not persons in the moral sense, and fetuses even less so, the Stoics felt it was appropriate (a duty) for parents to have a certain kind of attachment to their children. Namely, the kind of attachment which will encourage parents to raise their children to be as rational and disciplined as possible. But all things which are appropriate —

all duties — admit of exceptions, according to Stoicism. Thus it is generally
a duty to preserve one's own or another's (e.g., a child's) life. Yet the Stoics
did not judge life to be a good. In contrast, the natural order of things, and
acting in accordance with the perceived natural order, was a good. So if one
realizes that another's existence no longer contributes to the natural order of
things, reason requires that one ought to terminate that life. But, as I shall
emphasize in the next chapter when examining the Stoic position on suicide,
this imperative leads to serious difficulties. At best, the Stoic could claim to
know the natural order (fate) only in rough outline, not in sufficient detail to
determine precisely what role the life of any given individual plays in the
larger scheme of things. Therefore, deciding whether to commit infanticide,
say, could be extremely anguishing for the conscientious Stoic. Abortion
posed less of a moral ordeal since, *inter alia*, the fetus possessed no real moral
status to begin with. Since the fetus was not a member of the moral community, to abort violated no rule against homicide. True, Roman custom did
stipulate that in the case of a married woman, her husband's prior consent to
abortion was legally necessary.[77] But that the Stoics attached this legal
requirement to their moral theory is most doubtful.

The favorite argument of the Stoics in support of abortion was based on
an analogy. They held that the fetus is merely part of the mother's body,
analogous to the fruit of a tree. Therefore, technically she may do with the
fetus as she pleases until such time as it becomes ripe and falls down.[78] That
point was said to occur at birth when personhood in the biological sense,
as suggested above, was in actuality deemed to begin.

Thereafter, in the absence of overriding reasons which would require an
independent moral justification, the Stoics regarded the newborn as a future
member of the moral and legal community (though not a bona fide member).
As such, the newborn was usually treated as if he had the *prima facie* right to
life. Again, all things being equal, the newborn ought to be nurtured and
preserved. Hence the life of an infant must not be arbitrarily terminated on
the Stoic view. To be terminated, there must be a good, overriding reason for
doing so. In contrast, almost any reason was admissible for the justification
of abortion.

The main reason that was considered by the Stoics morally sufficient to
warrant infanticide was that the newborn was found to be defective. About
this Seneca is emphatic. He writes: "We destroy monstrous births, and we
also drown our children if they are born weakly and unnaturally formed."
He adds the justification that it is an act of reason thus to separate what is
useless from what is sound.[79]

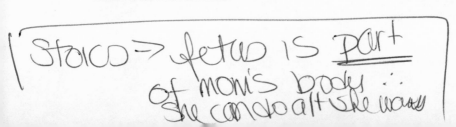
Stoics → fetus is part
of mom's body ..
she can do alt she wants

In my view, the full force of Seneca's argument is predicated on at least two aspects of Stoic ethics and physics. The first had to do with the eudaimonist emphasis on the importance of cultivating the required virtues in order to live well. Since, as I have suggested, the four cardinal virtues were said to be necessary to a life worth living, and since a malformed baby was, on account of its physical or mental defects, subject to being judged incapable of attaining the required virtues, to terminate the life of such defective newborns was seen as an "act of reason" from the Stoic perspective.

Second, this interpretation seems *a fortiori* all the more convincing in view of the Stoic teaching that whatever happens in this existence happens in accordance with Divine Providence. Hence there are no accidents and no chance events. Therefore, if a person feels regret over the workings of Nature — including the fact that some newborns are pathetically defective and are, in accordance with Divine Reason, allegedly better off dead — then this regret must be overcome by recalling that whatever happens is for the best. As Seneca himself taught, it is through human reason and its capacity to appreciate the harmony of the universe that individuals may soothe their sometimes aching spirits. For it is comforting knowledge that " . . . Divine Providence extends even to the smallest details of life."[80] If so, it is arguable from the Stoic perspective that to terminate a malformed infant is in fact to exercise a duty in accordance with God. For God (Nature) may be said to ordain infanticide as an act of human reason in conformity with divine justice on those occasions when we correctly grasp the difference between what is useless, and what is sound.

However, this Stoic position raises a definite criteriological problem concerning just how to tell when an infant is sufficiently defective to be judged useless. It also ignores the related difficulty of determining exactly in reference to whom the infant's life should be judged useful or useless. Therefore, these difficulties cast serious doubt on the soundness of Seneca's position on infanticide. The position on abortion is equally flawed: one can plausibly pressure the tenuous Stoic claim that the relationship of a pregnant woman to her fetus is merely analogous to that of a tree to its unripened fruit.[81] Persons and trees are such vastly different forms of life that it is only with incredible effort that this analogy can be made to stick. For these reasons, the above Stoic arguments in support of abortion and infanticide appear to be fraught with serious difficulties as they stand.

SUMMARY

To sum up our findings on the moral positions adopted by Pythagoras, Plato, Aristotle, and Seneca on the question of abortion and infanticide, Pythagoras alone was opposed to both of these measures. For the Pythagoreans, human life was said to begin at conception. It was a duty to God not to injure human or other sentient life. For bodily life, into which divine souls had descended as a punishment, possessed derivative value as the temporary residence in which souls could be purified by choosing to live wisely. Hence, to terminate either fetal or infant life was wrong because it deprived the victim of the immediate opportunity to at least attempt to live an embodied, earthly life in such a way as to win for his soul both the admiration and divine invitation of the gods. For this reason, the Pythagoreans would have also likely taken a dim view of contraception, through there is no evidence that it was formally condemned.

As for Plato, he was an advocate of both abortion and infanticide in the *Republic*. He saw in these biomedical measures the opportunity to promote the best possible offspring. Therefore, on utilitarian grounds predicated on his understanding of eugenic principles, he prescribed abortion and infanticide in specific circumstances that he judged to be useful to the overall quality of life in his utopia. These circumstances included the endorsement of infanticide for defective newborns or those born of parents considered past their prime for childbearing; and the endorsement of abortion in cases of adultery, pre-marital sex, or other unauthorized sexual unions which bore fruit. Although Plato does not address the question of contraception *per se*, he probably would have encouraged it for those couples who had exceeded his prescribed age for childbearing but were still sexually active. Since Platonists held that human life did not begin until birth, neither abortion nor contraception was considered to be morally objectionable.

As we saw, Aristotle, like Plato, also endorsed both abortion and infanticide on utilitarian grounds. But he put the start of actual human life at some point *in utero* when the fetus is informed by the rational soul, after first receiving at conception the nutritive and then, not long after, the sensitive souls. Aristotle in his *Politics* emphasized population control as a way to reduce poverty and civil unrest. But his argument in favor of infanticide is restricted to cases of defective infants alone. I argued that Aristotle's argument is ultimately grounded on his vision of the minimal good life and his implied judgment that highly defective newborns were better off dead in view of their permanent deficiencies, not to mention their possible drain on the resources of the well-functioning community. Furthermore, Aristotle considers abortion morally justifiable for population control purposes alone only when the fetus is incapable of sensation. But, as I argued, just when an individual fetus acquires a sensitive soul is very difficult to say with precision, using Aristotle's embryological guidelines. Therefore, the practical application of his moral theory on abortion appears flawed. I conjecture, in addition, that Aristotle might have ideally preferred contraception over abortion in the eligible case. If such measures had been effective (which they were not in his day), they would have limited population without raising the serious epistemological and practical problems of the sort described.

In contrast to Aristotle, the later Stoics considered human life in the biological sense to begin at birth with the child's first breath. It was not until the individual reached young adulthood (around the age of 14) that he became a person in the moral sense — with the full protection of the moral rule against homicide. But, like Aristotle, Seneca and the other Stoics prescribed infanticide for defective newborns alone on moral, if not on religious grounds. The defective newborn was taken to be so bereft of the opportunity to acquire the virtues necessary in order to live with some semblance of tranquility that it was considered an act of reason to acknowledge their unfitness by actively hastening their death. In addition, the Stoic position on abortion was more permissive than Aristotle's. The Stoics were on this question definitely opposed to the Pythagorean perspective which unconditionally protected fetal life. Their liberal approval of abortion even exceeded Plato's position in the *Republic*; unlike the Stoics, Plato would have at least required the prior consent of the state. But the Stoics likened the fetus to the fruit of the tree before it was ripe and has descended. Strictly speaking, the fetus possessed no moral worth of its own (since it had neither a *psyche*, nor rationality) and was ontologically akin to a mere plant. From this analogy, the Stoics argued that women had the discretionary right to abort at their own option. In so doing, they apparently chose to ignore the critical disanalogies to which their argument was also vulnerable.

But just as at life's beginning the human community faces questions of moral import, so too at life's end. To this latter group of questions we now turn in examining various ancient moral arguments for and against euthanasia.

CHAPTER SEVEN

THE PROBLEM OF EUTHANASIA

PRELIMINARY CONSIDERATIONS

Suicide and Language

In our prior examination of attitudes toward death, I argued that the Greek concept of euthanasia (a) referred to a manner of dying; (b) literally translated meant an easy or good death; and (c) logically entailed no one particular theory about what happened to the dead. This account must now be expanded in two important ways.

First, the modern reader, accustomed to restricting the current English versions of this concept to discussions of mercifully ending the life of a hopelessly suffering or defective patient, must appreciate that the Greek use of the term was broader in scope. The Greeks sometimes employed the term to describe *the spiritual state* of the dying person at the impending approach of death. Hence euthanasia was a term far broader in scope than contemporary English usage normally allows: its meaning was not anchored in medical contexts alone, though some of these contexts were also covered by it.

Second, for the Greeks euthanasia did not necessarily imply a means or method of causing or hastening death. However, when quick-acting and relatively painless drugs such as hemlock were first developed by the Greeks in the fifth century B.C., which allowed the individual to quit life in an efficient and bloodless manner, the linguistic result was that these forms of suicide were sometimes described as instances of euthanasia. But, even then, this ascription primarily involved a favorable appraisal of the subject's state of mind, not the means by which death came. Hence, on the whole there were no rigidly followed rules which fixed the application of the term euthanasia to contexts involving terminal illness or suicide alone, even though these situations could be properly referred to in selected cases.

Having thus recognized at the outset the decidedly broader scope which the concept of euthanasia enjoyed among the Greeks in comparison with the narrower contemporary English usage of this term, I propose in what follows to investigate one small piece of this Greek concept which comes closest to, and in fact originally included, what is often meant today by euthanasia. I

127

am referring to the notion of *voluntary* euthanasia. This is characteristically defined to mean that with the *prior consent of the patient*, either the patient or someone else may bring about a relatively quick and painless death to one who is judged on medical grounds to be the victim of an untreatable, painful terminal illness or disability. This limited definition of the term in question admittedly involves only one of the several eligible senses in which that term may be used in our own historical epoch.[1] But, despite this limitation, it does succeed in capturing two relevantly allied meanings which wed it to its older Greek ancestor: Namely, (1) voluntary euthanasia normally connotes a genuine human concern for the *psychological state of mind* of the suffering patient for whom life has apparently become an intolerable burden; *and* (2) it crucially links the *moral appraisal* of the conduct of the patient (and those supplying his medical services) to the precondition that *the patient is free to make a reasoned decision* regarding the option to hasten or not hasten his own death.

To clarify further, (2) above points to the fact that the Greeks did not admit *in*voluntary euthanasia as part of their general concept of euthanasia at all. If someone's life was terminated without his consent, normally this was *prima facie* a case of homicide. Such cases did not qualify as euthanasia principally because for adults, at least, it was popularly held that those who involuntarily died at the hands of others did not die well. Typically, they were not thought to have experienced good deaths. Just the opposite: if murdered, their souls were held to be restless, rancorous, and bent on revenge against their assailants.[2]

The obvious exception to all this was, of course, infanticide. But here the victim of infanticide was treated as a special case: a widely held Hellenistic view claimed that the infant was not really a human being until he was first fed.[3] Nor was pagan religion obsessed about the disposition of the deceased infant's soul, as Christianity was later to be. Hence, since infants were usually killed before they were fed, popular moral condemnation did not normally attach to this deed. As for those infants who *were* fed, though they were later involuntarily killed as an act of mercy or prudence owing to their chronic and painful ill-health, their deaths were customarily deemed tragic and pitiable. Such deaths were not, in the Greek mind, standardly considered good or easy deaths either for the victims or for the parents. At best, they were regarded as necessary. Perhaps reflecting this fact, popular Greco-Roman sentiment tended to resist the killing of *non-defective*, healthy infants even though Roman law did not prohibit this practice until the fourth century A.D., as we have already observed. Therefore infanticide, which is by its very nature involuntary since the infant lacks the verbal skills by which to give or withhold consent, was not customarily included in the Greek concept of euthanasia.

Moreover, (1) above cannot be overemphasized. When talking about voluntary euthanasia, the current tendency is to freight the concept heavily with its moral, appraising sense. But in so doing, this misses the dominant ancient connotation of this term which *primarily* involved a psychological appraisal, and only secondarily a moral one. To illustrate this subtle difference, I suspect that the pivotal ancient question regarding euthanasia in its broadest sense was, typically: Did the subject voluntarily meet death with peace of mind and minimal pain? In contrast, the looming contemporary question involving euthanasia tends to be: Is euthanasia under any conditions morally justifiable?[4] Moreover, I do not think it is correct to infer from this that the Greeks were oblivious to the moral challenge voluntary euthanasia represented. They did not ignore the task of trying to furnish a coherent ethical defense of the practice. In fact, after we explore what some of the Greek and Roman philosophers had to say about the general question of the morality of suicide, I think quite the opposite impression will be left. At any rate, I do contend that the term's ability to confer or withhold a favorable appraisal of the apparent state of the subject's mind during his final conscious days and hours marked its primary linguistic function. Despite this emphasis, however, it is doubtless true that such a psychological appraisal was not readily separable from the allied moral appraisal of whether the subject evinced sound thinking, on moral or other grounds, for choosing to withdraw from life. Hence the Greek *euthanasia* was indeed capable of bearing two senses, both psychological and moral, and surely in everyday discourse these two senses were sometimes mingled together.

As to the related notion of suicide by which I here mean, in a rough and preliminary way, those forms of conduct in which the individual intentionally kills himself, the Greeks and Romans were generally sympathetic provided that the deed was done *for the right sorts of reasons*. Just what these reasons amounted to from the vantage point of some of the leading philosophical schools is a question which awaits our further treatment. Presently, it is worth noting that of all the ancient cultures it is in the Greek culture of the fifth and fourth centuries B.C. that suicide is first represented as a *mode of dying* instead of as an act of killing, which was its historically prior sense.[5] Why this is so can only be conjectured. But the philologist David Daube plausibly suggests that in the case of suicide, " ... the most conspicuous thing for the onlooker is that a man kills himself: it is this arresting deed that the earliest words notice. That he dies in a certain fashion is a more subjective feature, a feature concerning him alone, to which attention is paid only at a later, more reflective stage."[6] It was especially due to the flowering of Greek philosophy and, I would add, to the peculiar emphasis of eudaimonist ethics which tended to link isolated appraisals of acts of personal conduct to more general appraisals of the moral agent's whole way of life — from birth to approaching death — that likely fostered this higher reflective stage with respect to acts of suicide. For a person's life was viewed as a whole; and hence his birth, life, and death were open to moral as well as aesthetic evaluations based on considerations of proportion, measure, and order. To end life at the right time, then, approached an *aesthetic* challenge for some.

Furthermore, a second factor may have been the introduction and perfected use of hemlock. As a matter of fact, this agreeable potion reportedly did contribute to an increase in the suicide rate in antiquity.[7] Daube states:

Its role in the human carrying out of capital sentences is familiar from the story of Socrates; and it may well have been a factor in the creation of such myths as that of a utopian island whose inhabitants, when the desirable maximum span of life is reached, take their leave by lying down on a plant that produces eternal sleep.

It would also make for a linguistic turn toward a more passive concept of suicide — as a mode of dying rather than killing. When a man falls upon his sword, the external violence, *the killing*, is incomparably more impressive than when he takes a drink that brings about a slow ebbing away of his life — *a dying*.[8]

One can add to this the additional point that insofar as what we now call "voluntary euthanasia" was construed by the Greeks and Romans as one possible type of suicide, the phrase bears a definite family resemblance to the Latin expression for suicide, *mors voluntaria*, which also stressed the voluntary nature of such dying. In contrast, our English word "suicide," which literally means self-killing, lays stress on the peculiar fact that oneself is the object of one's killing. In this, the English language has lost touch with many of the most characteristic Greek and Roman expressions for suicide which tended to emphasize suicide as a mode of death, and so decidedly downplayed the novel and sometimes violent side of self-killing. It is precisely this sort of linguistic legacy, incidentally, which in my view lends support to the claims of Joseph Margolis. Margolis has argued that, given the culturally variable character of the ascription of suicide, there can be no truly value-neutral concept of suicide. What is allowed or disallowed linguistically is always colored by some specific partisan perspective which is adopted (consciously or unconsciously) by the individual observer.[9]

Indeed, even within a single culture one may discover a rich and variable repertoire of expressions which refer to suicide. This is especially evident among the Greeks, whose historians, playwrights, and philosophers coined many such expressions. To sample just a few of these, Daube records Xenophon's *haireo thanaton* (to seize death); Euripides' *aporregnymi bion* (to break off life); Plato's *hekonein haidou* (to go voluntarily to Hades); Aristotle's *biazesthai heauton* (to flee living); and the Stoic's *eulogos exagoge* (sensible removal).[10] Surprisingly, the first noun phrase for suicide does not regularly appear in Greek until the first Pre-Christian century. It was *autocheiria*, "an act with one's own hand," or "own-handedness." Somewhat later, in the second century A.D., *autoktonia*, "self-killing," became standard.[11] But by that time the concept of suicide as *a way of dying*, rather than as a peculiar form of killing, was thoroughly embedded in Greco-Roman thought. Hence such thinking regularly tended to overlook the literalness of these later noun phrases.

We must turn now to briefly consider two remaining points. First, the types of suicide recognized by the Greeks. And second, the typical methods by which this act was carried out. Following that, we shall examine some of the leading philosophical schools to discover whether they would have supported or opposed that form of suicide which we call *voluntary euthanasia*, and which the Greeks, too, would have thought of in that way.

Types and Methods

It is possible to distinguish at least eight recurring types of suicide that were recognized as such during the Greco-Roman period.[12] What distinguishes these from each other is primarily a variable estimate of the supposed motives of the subject for taking leave from this world. My purpose here is to suggest where in this constellation of ancient specimen cases having to do with self-murder the concept of voluntary euthanasia would have likely belonged.

It must be added that I shall limit my account to non-compulsory instances of suicide. It is true that the ancients sometimes forced people to take fatal poisons in capital executions. It is also true that they occasionally recognized cases of suicide that were thought to be linked to some sort of mental health disorder.[13] But the former case is of little concern to us here, since suicide by direct coercion precludes the relevant sort of moral appraisal of the praiseworthiness or blameworthiness of the agent's conduct. That is, one would surely not be inclined to praise or blame the conduct of one who had no other real choice but to imbibe a fatal poison; the question may even be legitimately raised as to whether this is properly speaking an instance of suicide at all. Similarly those individuals who, on account of some form of illness or disability, are judged to have lost control of their senses and conduct are thereby also rendered ineligible subjects of moral appraisal. In antiquity, as now, it was presumed that individuals so qualified were in some sense acting involuntarily.[14] But it is suicide in the Latin sense of a voluntary death (*mors voluntaria*) that deserves our utmost attention here. For we are preparing to analyze ancient philosophical arguments for and against non-compulsory instances of this deed. And these arguments ultimately presuppose some capacity for reason and choice on the part of the individual.

Resuming our main task, at least eight types of non-compulsory suicide, distinguishable primarily by their presumed leading motives, are discernible in Greek and Roman literature. These may be conveniently divided into two somewhat arbitrary groups. First, *heroic suicide*, characterized by the subject's desire to escape present or impending shame and dishonor; or the noble wish to devote one's life to one's country; or to slay oneself in order to relieve excessive grief. Second, *non-heroic suicide*, which, by degrees only, tended not as a group to be as inspiring to the ancients, perhaps partly because they were perceived to be on the whole more egoistically motivated or because they were thought to occur regularly in less glorious circumstances.

To the first *heroic* group belong four recurring cases motivated by: (1) the desire to escape shame and dishonor, as in the case of Ajax, who, after realizing that he had slaughtered his own camp's herds in a fit of madness, fell on his sword;[15] (2) the desire to demonstrate total self-devotion to one's country, as in the case of the last Athenian king Kodras, who, after learning from the oracle at Delphi that the only way his country would not be captured by the Spartans was for him to die at their hands, arranged in that manner for his own self-sacrifice;[16] (3) the desire to end unrelieved grief over the death of a loved one, as in the sad tale of King Aegeus, who, owing to his receipt of the false signal that his son Theseus had perished, threw himself into the sea which today bears his name;[17] and (4) the desire to atone for the death one has accidentally caused to another person, as in the case of Adrastos, who, having inadvertently killed the son of Croesus, begs Croesus to arrange for his own death in recompense.[18]

To the second *non-heroic* group belong four additional types of suicide motivated by: (5) the desire to end personal grief owing to persistent and degrading poverty, which was explicitly recognized in the writings of Theogonis as sufficient cause to quit life;[19] (6) the desire to be reunited in death with a deceased loved one or companion, as in the case of Paris' beloved Oenone, who, owing to her intense longing and sorrow, threw herself on his funeral pyre;[20] (7) the desire to quit life owing to the sheer sense of how boring, pointless, and futile it may appear, a motive which, significantly, is often redescribed as a form of madness since most bystanders are not so powerfully in agreement, and one which may have been at work in the mass suicides of the maidens of Miletos;[21] finally, (8) the desire to escape a relentlessly painful illness or infirmity, usually thought to be incurable in nature, and usually (though not always) associated with the approach of old age. This last type of suicide, which we may call voluntary euthanasia, is illustrated in the case of the elderly citizens of Ceos, who, once they had passed age 60, were in the habit of taking leave from life by drinking hemlock.[22] Also, there is the case of Titus Aristo, who, not untypically by the period of the Roman Empire, carefully reasoned that he would end his life if his physicians reported that his insufferable illness was definitely beyond their means to cure.[23]

In sum, it is to this second group of non-heroic suicides that voluntary euthanasia in most cases belonged. The last illustration of the suffering patient, Aristo, and our preceding observations about the possession of lethal drugs by the ancient physician point to the unique opportunity of the doctor to declare his patient untreatable, and, if so requested, supply to him the appropriate "cup of death." Since, especially by the late Hellenistic and Roman Imperial Period, almost all of the older religious stigmas attaching to suicide had lost their force and since, excepting the cases of slaves and soldiers, civil penalties against suicide were practically unknown throughout both Greece and Rome, there was little to constrain an individual bent on quitting life from making good his resolve.[24] The methods chosen are readily imaginable. They included, besides the ever-popular hemlock, self-immolation, jumping from high places, drowning, self-starvation, hanging, and running oneself through by the sword or dagger. Moreover, of these foregoing methods, starvation alone would qualify in modern times as *passive* voluntary euthanasia, wherein the subject's death is brought about by *omitting* those elements necessary to the support of biological life. The others would qualify as *active* forms of euthanasia, wherein death occurs as the more immediate result of *committing* a fatal act of some sort.[25]

All this prompts one to ask: On what moral grounds did some of the leading ancient philosophers support or oppose voluntary euthanasia (either active or passive) in the sense we have characterized it above? In particular, what position did the Pythagoreans, Plato, Aristotle, and the Stoic Seneca take on this issue, and what bearing, if any, did their attitude toward death likely have on their moral or religious reasoning in regard to suicide or euthanasia?

PHILOSOPHICAL PERSPECTIVES

Pythagoras

It is not necessary to repeat here the tenets of Pythagoras already discussed in the preceding chapter concerning the doctrines of (a) _divine personal immortality_ of the soul; (b) transmigration of souls upon death; (c) earthly life as penal servitude marked by the fallen soul's entombment in the body; and (d) earthly life as an opportunity to purify the soul through the proper study of philosophy and mathematics, so that one day the soul might be liberated and return to its origins in heaven. Instead, let us focus on the fact that the Pythagorean opposition to all forms of suicide, including euthanasia, ultimately rests on a religious principle from which this moral teaching is derived. In our discussion of abortion, I alluded to the Pythagorean religious principle that it is a direct violation of an individual's higher duty to God to behave in such a way as to prematurely end his or another's life. In the _Phaedo_, this principle appears to be linked by Socrates to the teachings of the fifth century B.C. Pythagorean Philolaus:

> The allegory which the mystics tell us – that we men are put in a sort of guard post, from which one must not release oneself or run away – seems to me to be a high doctrine _with difficult implications_. All the same, Cebes, I believe that this much is true: that the gods are our keepers, and we men are one of their possessions.[26]

The religious character of the absolute Pythagorean rejection of suicide is readily discernible. The entire argument presupposes the crucial premises that: (1) God exists; (2) man's highest moral duty is always to obey the commands of God; and (3) one of these divine commands unconditionally bars the individual from taking early leave from his embodied, earthly existence. Therefore, if any one of these three premises is in turn successfully undermined by the religious skeptic, the Pythagorean position against suicide at once collapses. Perhaps it is in part Plato's awareness of the inherent frailty of this religiously endowed argument that led him to make Socrates acknowledge the "difficult implications" of this perspective. For it is a perspective founded ultimately on mysticism and faith – not reason and dialectic – which Plato and Socrates both much preferred, notwithstanding their own admitted religious tendencies.

Before leaving the Pythagoreans, I would like to make three related points having to do with my earlier claim that theirs was a philosophy imbued with the respect for human life ethic.

First, the word "life" is, as already noted, ambiguous.[27] One can have respect for: (a) each individual human life; (b) the life of the human species taken as a whole; (c) the life of one's lineage; and (d) the life of one's tribe, nation, or race. Frankena is right in saying that it is primarily concern for (a), the individual's life, that is of central concern in bioethical discussions of euthanasia and abortion.[28] He is equally right, I think, in observing that, in speaking of the individual, we can speak of his bodily or physical life; and, in addition, we can speak of his mental or spiritual life. The point is that on the Pythagorean view, it is the *spiritual* life that has the highest value owing to its divine religious character. Therefore, if bodily life has value, as it certainly does for the Pythagoreans, it is wholly parasitic on the value it receives from being the temporary residence of the divine soul. Even so, in our ordinary conversations about the value of human life, it is really *bodily* life that, for all intents and purposes, is minimally at issue. For it is most clearly life of this physical sort that is prevented or terminated in acts of abortion, infanticide, suicide, euthanasia, and the like.[29]

Second, to this must be added the distinction between respect for human life in the biological, physical sense just defined (*zoa*), and respect for the career and personality of the individual (*bios*). These two senses are by no means equivalent, as witnessed by the fact that one could assert without contradiction that, although Jones possessed life in the first sense, he was entirely devoid of life in the second sense following his unfortunate automobile accident. Hence, the alleged sanctity of human biological life, and the sanctity of human personality, bear at most a loose logical relation. If so, this implies, against the Pythagorean position, that one can fairly claim to be a partisan of the respect for human life ethic, in the primary sense (a) taken above, and still legitimately admit overriding conditions for the moral justification of euthanasia.

Third, it is revealing that throughout human history there have been few partisans who, without any qualification at all, have endorsed the respect for human life ethic in and of itself. For if one were such a partisan, then one would have to be willing to assert that human bodily life has absolute value *per se* simply because it is human life. That claim would be taken as a *basic*, unappealable first principle. Yet even the Pythagoreans did not choose to make this basic claim: nor, incidentally, did the Christians of later antiquity.[30] For both groups found little satisfaction in such a bold, tautologous, and empty-sounding secular principle. And so, although from vastly different religious traditions, Pythagoreans and Christians alike made the value of human bodily life at least partly if not totally conditional on the higher and antecedent principle that God valued man. Therefore, it remains true that the Pythagoreans were partisans of the respect for human life ethic. But, even for them, it is questionable whether human life was taken to be intrinsically valuable *per se*. Its value, though genuine and of binding importance for their medical ethics, was a *derivative* value resting ultimately on religious, not moral, considerations.

Plato

If one assumes that the opposition to suicide voiced by Socrates in the above quoted *Phaedo* passage (p. 134) mirrors the actual sentiments of Plato, and if one further takes into consideration the discussion of euthanasia for the incurably sick in passages from *Republic* and *Laws*, one senses an evolution taking place in Plato's thinking on this subject. It seems to me that the absolute opposition toward suicide voiced at *Phaedo* 62b, ff. is significantly softened at *Republic* III. 405, ff. This is done in such a way as to furnish there a moral argument in favor of rational suicide among the incurably ill. Moreover, at *Laws* XI, 873 c, ff., which was probably written when Plato was an old man himself approaching death, it appears that a decidedly intermediate position between the two earlier accounts is struck. The account finally given in the *Laws* both moderates and seeks to incorporate salient features of Plato's earlier antithetical positions.

Of the three relevant passages, scholars have devoted more attention to the *Phaedo* than to the other two combined. This is certainly understandable. The *Phaedo*, together with the *Apology* and *Crito*, includes the earliest sustained treatment of the timeless moral and legal questions associated with death and dying ever to have been bequeathed by antiquity to the modern era. A fair amount of attention has thus been paid to the problem of attempting to unravel which views in the trilogy were owned by the historical Socrates, and which by his student and admirer Plato. I shall not stop here, however, to enter into that particular debate. But I shall venture to express my sympathy for those commentators who have linked the historical Socrates with the comparatively agnostic view toward death expressed by the figure of Socrates in the *Apology*. There, at *Apology* 40C, ff., the following alternatives are advanced:

Death is one of two things. *Either* it is annihilation, and the dead have no consciousness of anything, *or*, as we are told, it is really a change – a migration of the soul from this place to another. Now if there is no consciousness but only a dreamless sleep, death must be a marvelous gain ... If on the other hand death is a removal from here to some other place, and if what we are told is true, that all the dead are there, what greater blessing could there be than this, gentlemen?[31]

Significantly, although Socrates does not here claim to know which account of being dead is correct, he does claim that in either case death is nothing to fear. What's more, this was an original insight, the importance of which was not lost on the Epicureans and Stoics who followed in the Socratic tradition of debunking the terrors of the afterworld.

If we recast Socrates' preceding disjunction into the terminology suggested in Chapter Three of this book, his position comes close to saying that at death one faces either the prospect of (1) *natural personal dissolution* or (2) a variant of *chthonic personal immortality*. (The latter he strives to dress up in order to emphasize the potential for pleasantries even in Hades!) The above digression into the question of what happens when one dies appears to get settled in the *Phaedo*. For there Socrates is made to embrace the second disjunct: the soul is immortal. Moreover, the immortality it possesses is depicted not with Homeric, chthonic imagery, but rather with the Orphic-Pythagorean imagery associated with *divine personal immortality*. Furthermore, life itself is now described as a preparation for death. The true philosopher does not recoil with fear at the thought of his own death but anticipates it with longing. For death, the separation of the soul from the body, at last affords him the ultimate opportunity to behold universal knowledge in its purest splendor, unencumbered by the mere appearances and base distractions with which he had formerly been afflicted while embodied on earth.

All this brings us to a central puzzle regarding the moral acceptability of suicide in the *Phaedo*. Namely, if it is conceded for discussion's sake that the above account in the *Phaedo* represents a true picture of the afterworld, is it not then natural to ask why philosophers, at least – after their souls have been sufficiently purified – ought not take early leave from their inferior bodies by suiciding? To this question Socrates unequivocally replies at *Phaedo* 62C, ff.: "No man has the right to take his own life, but he must wait until God sends some necessity upon him, as he has now sent to me."[32] Socrates' reasoning here also appears to be grounded on religious premises. It bears a faint resemblance to the Pythagorean metaphor (p. 134) of the soul being required to serve a faithful guard duty for and in the body, as a form of penance from which " . . . one must not release oneself or run away."

But, on closer inspection, Socrates' argument is not quite the same is the Pythagorean's. First of all, it is conjectural whether Socrates or Plato embraced the Orphic-Pythagorean view that embodied life is a penal servitude for past sins. It was not central to Socrates' teaching, nor was it central to Plato's, that human life was a form of punishment. Instead, both men acknowledged the mixed character of human existence, and both emphasized the value of living a reflective philosophical life in order to successfully meet the challenges of this middling, earthly plane. Second, the Pythagoreans, emphasizing the penance constituted by human life as a duty placed on man by the gods, argued that humans therefore had no right to end their penance by committing suicide.

This is not, however, the content of the full argument developed by Socrates at *Phaedo* 62 b–c, ff. Rather, Socrates is made to rest his case against suicide on the analogy that " . . . the gods are our keepers, and we men are one of their possessions." Moreover, he suggests that the gods are very good masters and that, as their mere possessions or chattel, it would be morally wrong for individuals to take flight from their rightful and benevolent masters through an act of suicide, barring some divine necessity or compulsion (*ananke*). Such a compulsion would presumably need to be an *external* one, like Socrates' predicament of facing capital punishment via poisoning. Hence, quitting life out of a sense of boredom, out of personal fear, out of a desire to escape earthly struggles, or out of thoughtlessly succumbing to destructive internal impulses, would in no case pass muster morally. Therefore, the philosopher who is motivated to commit suicide by the desire to win immediate universal knowledge (and in so doing escape the earthly struggles of inquiry), would be violating his higher duty to the gods. The true philosopher finds life to be a "rehearsal" for death only in a special sense. By living in a reflective manner so as to carefully prepare his immortal soul for its eventual release from the body due to natural causes, he may succeed in conditioning his real self – his soul – for its most inspiring and satisfying odyssey. He does not, like a greedy guest, rush uninvited to death's banquet; nor, like a disobedient slave, does he willfully desert his benevolent keeper. Such is the negative picture of suicide we get in the *Phaedo*.

The *Republic*, however, appears to all but ignore these religious constraints against suicide. In the *Republic* Socrates is made to advance the argument that *rational suicide* is not only possible but morally warranted in at least two related cases. The whole issue of whether suicide can cogently be said to be rational is a hotly debated one in our own time. By ascribing such a position to *Republic* III. 405, ff., I here mean that Socrates talks as if sound reasons do in fact morally justify suicide in cases of (1) incurably debilitating disease or (2) incurably debilitating disablement. If so, for our purposes Plato may be said to support *voluntary* euthanasia for adults in the *Republic*. By extension, *involuntary* euthanasia may also here be understood to be justified on similar grounds for incurably defective younger children. This latter inference would then further serve to illuminate Plato's qualified support of infanticide for defective newborns — a position of his found to be somewhat flawed in our earlier discussion.[33]

What, then, are his grounds for supporting euthanasia? They appear to be primarily utilitarian, not religious, in nature. Speaking in admiring tones of the legendary physician-god Asclepius, whom he compares favorably over contemporary practitioners like Herodicus, who practice bad medicine by coddling and nursing their hopeless patients (and themselves) to a lingering and painful death, Socrates asks:

Shall we say, then, that Asclepius recognized this [folly] and revealed the art of medicine for the benefit of people of sound constitution who normally led a healthy life, but had contracted some definite ailment? He would rid them of their disorders by means of drugs or the knife and tell them to go on living as usual, so as not to impair their usefulness as citizens. But where the body was diseased through and through, he would not try, by nicely calculated evaluations and doses, to prolong a miserable existence and let his patient beget children who were likely to be as sickly as himself. Treatment, he thought, would be wasted on a man who could not live in his ordinary round of duties and was consequently *useless to himself and to society*.[34]

The gist of Plato's position, in effect, is this. In his ideal society, he envisions the welfare of the state as a whole to hold a higher worth than the welfare of any one individual. Duties to the state here override duties to self and to family. In fact, religious duties, one recalls, are in Plato's utopian scheme finally invented and orchestrated to serve the purposes of the state as "convenient fictions." His aim in so doing is to further influence the masses so that they make whatever personal sacrifices are deemed necessary in order to insure the full productivity and strength of the community. Hence, Plato's support of euthanasia in the *Republic* drops the religious underpinnings and analogies found in the *Phaedo*. Dropped, too, is an explicit eschatological argument attempting to allay natural human fears of the unknown that are associated with death. Substituted in their place is the more immediate specter of living out a tormented existence. This last prospect, made possible by a person's jealously guarded option and allegedly selfish insistence on surviving as a sickly, parasitic, unproductive citizen, is vigorously ridiculed as morally blameworthy behavior.[35]

It follows that hopelessly sick individuals are to a very real extent expendable from the moral point of view, according to Plato's *Republic*. Thus it is safe to infer that in the case of older children and adults, he approved of *voluntary* euthanasia on utilitarian grounds as an eligible form of rational suicide in the aforementioned instances. Whether he also approved *involuntary* euthanasia in the case of adults remains an open question which he does not speak to explicitly in the *Republic*. Moreover, it would be unfair, I think, to convict him of this on his silence. But that he approved involuntary euthanasia in the case of infants there can be no doubt. Though Plato acknowledged infants as human beings, he regarded their continued existence, if incurably sick or defective, as an unreasonable liability too great for the state to endure. Unlike the Pythagoreans, who in other ways informed much of Plato's thinking, their respect for human life ethic was one moral principle he did not care to own in their absolute terms.

Yet, oddly enough, in Plato's last major work, *The Laws*, which, like his earlier *Republic* shares the goal of imagining an ideal commonwealth in which to live, Plato condemns suicide as morally wrong in all but four cases. He further prescribes legal penalties according to which the bodies of suicides must be (a) buried alone apart from all others; (b) buried "in disgrace" in deserted places which have no name, somewhere on the outskirts of the state; and (c) buried without any headstone or marker. Doubtless these sanctions reflect in part some of the traditional religious stigmas associated with blood pollution, and specifically pollution by self-murder, still current in some of the city-states of Plato's own day.[36] More important for us, however, is the observation that of the four excusing conditions Plato lists as *not* deserving of moral blame, the second explicitly excuses individuals, like Socrates in the *Phaedo*, who take their lives in obedience to a legal order from the state. No less important, the first and third excusing conditions could apply to those cases of the incurably ill or disabled, like Plato refers to in the *Republic*, who are deemed morally justified for electing voluntary euthanasia. (The remaining fourth condition excuses those who slay themselves in order to atone for "some irremediable disgrace" with which they cannot live.)[37]

The one remaining motivation for suicide which was the most prevalent in antiquity, and, I dare say, remains so in our own day, is described in the *Laws* in condemnatory terms as follows. It is the case of one who " . . .

imposes this unjust judgment on himself in a spirit of slothful and abject cowardice."[38] In other words, it is the case which covers all those (including the pseudophilosopher) who, acting from the mere motives of boredom, sloth, fatigue, or simple rejection, take it upon themselves to impulsively quit life. This is most clearly the kind of suicide opposed by Socrates in the *Phaedo*. It is also the kind which Plato studiously avoids endorsing in the *Republic*. Yet the case of voluntary euthanasia, which Plato does endorse there on moral grounds of a utilitarian sort, is also specifically excused (if not endorsed) in the *Laws*. In unmistakable language to that effect in condition three, alluded to above, Plato's proposed punitive legislation does *not* apply to those whose hand is ". . . forced by the pressure of some excruciating and unavoidable misfortune."[39] For such are the misfortunes of incurable injury or disease, to be sure.

Hence, there is little doubt in my mind that in his old age Plato sought to harmonize in some condensed fashion his earlier thoughts on the topics of suicide and euthanasia as found in the *Phaedo* and *Republic*. But whether or not he struck a morally defensible compromise in his *Laws* is left, characteristically, for his readers to decide.

Aristotle

One who disagreed with Plato's stand in support of euthanasia was his student and former colleague at the Academy, Aristotle. It seems that Aristotle did not devote so much of his attention to the problem of suicide as did Plato. But what he contributes to the discussion in two key passages from his *Nicomachean Ethics* is both concise and to the point. What these two passages suggest is his unqualified opposition to suicide, and, in particular, his opposition to the proposition that it is morally acceptable for a person suffering from an incurable disease or disability to quit life.[40]

This opposition to suicide is expressed in one of these two passages in the context of his broader examination of the question of whether a person can be properly said to treat himself unjustly.

> . . . one class of just acts are those acts in accordance with any virtue which are prescribed by the law; e.g., the law does not expressly permit suicide, and what it does not expressly permit it forbids. Again, when a man in violation of the law harms another (otherwise than in retaliation) voluntarily, he acts unjustly, and a voluntary agent is one who knows both the person he is affecting by his action, and the instrument he is using; and he who through anger voluntarily stabs himself does this contrary to the right rule of life, and this the law does not allow; therefore he is acting unjustly. But towards

whom? Surely towards the state, not towards himself. For he suffers voluntarily, but no one is voluntarily treated unjustly. This is also the reason why the state punishes; a certain loss of civil rights attaches to the man who destroys himself, on the ground that he is treating the state unjustly.[41]

The point is that the person who suicides is blameworthy not just because he has seen fit to mortally injure himself (which in itself is deplorable as being " . . . contrary to the right rule of life," i.e., unnatural). Most of all, the person who suicides is blameworthy because his act violates his duty to the state to remain a productive, contributing citizen so long as he possibly may.

This raises two questions. First, is Aristotle really intending to include here the incurably ill, who by their disabilities may already be unable to render useful service to the state? Second, what, in any case, is the logic behind punishing the *successful* suicide with "a certain loss of civil rights"? Let us pursue these questions in turn.

It will be recalled from our earlier examination of Aristotle's attitude toward death that he squarely labels death ". . . the most terrible of all things; for it is the end, and nothing is thought to be any longer either good or bad for the dead."[42] Moreover, in his writings on courage, he defines the brave man as " . . . fearless in the face of noble death" [i.e., death in battle], and fearless as well in " . . . all emergencies that involve death." The admirable courage of the brave man confronted by the personal prospect of death or disease stems almost entirely from his sense of what is honorable conduct and what is base, says Aristotle. This rests, ultimately, on his intellectual ability to find the mean relative to himself between the opposing vices of cowardice and rashness. He may grasp some notion of this mean by observing the conduct of the prudent man who has mastered the right moral rules by which ends are achieved. Moreover, psychologically even the courageous person, according to Aristotle, is not invulnerable to fear at all times: there are some things deserving of fear. For example, ". . . things terrible even beyond human strength."[43] Yet, in general, even such things as these (and certainly impending death and some forms of painful disease would qualify) are only feared by the brave person ". . . from the right motive, in the right way, and at the right time. . . ."[44]

It is fair to inquire how the virtuous person ought, in Aristotle's view, to conduct himself if faced with a terminal, debilitating disease which saps his powers to exercise the minimal duties of citizenship. Would such a person be excused from moral censure if he killed himself? The answer is no. On this Aristotle appears to be unequivocal:

But to die to escape from poverty or love *or anything that is painful* is not the mark of a brave man, but rather of a coward; for it is softness to fly from what is troublesome, and such a man endures death not because it is noble but to fly from evil.[45]

Thus, I think it is safe to conclude that Aristotle did not approve of the individual who quit life on account of the infirmities of illness. And I further conclude that Edelstein, who attempts to argue that Aristotle's opposition to suicide did not really involve moral censure but only amounted to a plea for more backbone, is just wrong about this.[46] Furthermore, our evidence suggests that euthanasia was blameworthy for Aristotle not simply because it deprived the state prematurely of one of its own. It was blameworthy as well because such an act typically constituted an excessive degree of rashness or cowardice. This was especially true of the subject who ended life out of excessive anger or grief at his helpless suffering. Or the subject who acted deficiently by not meeting his predicament with the requisite courage called for in order to face nobly the trial of his affliction.

Secondly, as to the logic of calling for civil penalties involving the loss of rights for the successful suicide, surely some sense can be made of this. In the context of the city-state, Aristotle is particularly intent on condemning the act of suicide, not simply the actor. Although he does not specify what this loss of rights involves, I conjecture that he has in mind special prohibitions concerning the burial of suicides similar to those prohibitions mentioned by Plato in *The Laws*. For example, it is known that in Aristotle's day the hand of the suicide was sometimes severed and buried separately from the remaining corpse. Moreover, the institution of nameless graves, and the prospect of inglorious and isolated burials far from the hallowed burial grounds of the community's honored dead may have served a double purpose as a warning to the living who were considering suicide or euthanasia.

A final question. Since, as I have argued, Aristotle opposed both suicide and euthanasia on moral grounds without any apparent exception, can he be identified as a partisan of the respect for human life ethic in either its basic or derived sense?[47] I think not. For we saw that to be a partisan of the respect for life ethic, one must either accept as basic the proposition that human bodily life has absolute value *per se* just because it is human life. Or one must accept a similar proposition that would also confer absolute value to human life even though that value derives from prior principles. In either case, Aristotle fails to qualify. It is true that he opposes suicide. But his moral justification of infanticide for defective newborns on utilitarian grounds disqualifies him from further serious consideration. For is it not self-contradictory to assert, on the one hand, that all human bodily life has absolute value just because it is human; and, on the other hand, that some human lives should be terminated just because they are judged to be defective? Such a position would appear inconsistent. It represents, at any rate, a peculiar departure from our ordinary way of talking about respecting the lives of our fellow human beings.

Seneca

We have seen that the Stoics viewed death as nothing to fear. In the main, they taught a version of *natural personal dissolution* according to which the tiny fiery pneuma indwelling within each person dispersed at the moment of death to reunite with the world-soul Pneuma of God.[48] Thus there were no punishments awaiting the deceased's ghost in the netherworld, for there were no ghosts. Individuality and personality effectively ceased with death. Death was viewed as the natural and inevitable resolution of human life itself. Hence only the fool would attempt to oppose death; the wise man would arrange to meet it head on if he could, timing his exit from this world with the aplomb of a thoughtful dinner guest who seeks to avoid boring or detaining his hosts. Hence the wise man withdraws from life's banquet in just the right way, under just the right conditions.[49]

This raises two important questions. First, did the Stoics hold that suicide, and particularly euthanasia, was a rational act? Second, did they hold that each person possessed what amounts to a moral right to end his own life? I shall argue that the writings of Seneca suggest affirmative answers to both of these queries. But, as we shall see, they do so only by insisting on some very strict qualifications that were quite characteristic of the Stoic perspective on suicide and euthanasia.

In general, the Stoics were not the first philosophical school to approve of suicide. We have already seen that Plato approved of it in a limited way, especially for the incurably ill who suffered their fate under compulsion (by external necessity, as it were). Moreover, representatives of two minor Socratic schools, the Cynics and the Cyrenaics, at various times also looked favorably upon suicide.[50] What distinguished the Stoics, however, was that they were mightily inspired by the moral example of Socrates' courageous death. Following Socrates, they viewed philosophy not only as the art of right living, but also as the art of dying well. Very likely, learning to die nobly became one of the central problems in Stoic ethics partly because the Stoics came to view the act of dying as an important test of a person's moral strength and stature.[51] Thus it is little wonder that the Stoic founder

Zeno committed suicide in his old age prompted by the agonizing pain of a foot injury; his successor, Cleanthenes, suicided after developing a boil in his mouth; Seneca himself slashed his wrists at around age 68 after it appeared certain that his former student, Nero, would otherwise order him executed on the dubious charge of treason.[52]

This last mention of the circumstances of Seneca's death, in particular, highlights a crucial Stoic idea concerning suicide: to have the capacity to willfully depart from life was viewed as the greatest boon to mankind; it constituted a real choice that epitomized moral freedom. In the following passage, Seneca expresses this central idea. He also tacitly rebukes thinkers like Pythagoras and Aristotle who opposed suicide, and demonstrates his sympathy for voluntary euthanasia on account of the cruelties of disease.

You can find men who have gone so far as to profess wisdom and yet maintain that one should not offer violence to one's own life, and hold it accursed for a man to be the means of his own destruction; we should wait, say they, for the end decreed by nature. But one who says this does not see that he is shutting off *the path to freedom. The best thing* which eternal law ever ordained was that it allowed to us one entrance into life, but many exits. *Must I await the cruelty either of disease or of man*, when I can depart through the midst of torture, and shake off my troubles? *This is the one reason why we cannot complain of life*: it keeps no one against his will. Humanity is well situated, because no man is unhappy except by his own fault. Live, if you so desire; if not, you may return to the place whence you came. You have often been cupped in order to relieve headaches. You have had veins cut for the purpose of reducing your weight. If you would pierce your heart, a gaping wound is not necessary; a lancet will open the way *to that great freedom*, and tranquillity can be purchased at the cost of a pin-prick.

What, then, is it which makes us lazy and sluggish? None of us reflects that some day he must depart from this house of life; just so old tenants are kept from moving by fondness for a particular place and by custom, even in spite of ill-treatment. Would you be free from the restraint of your body? Live in it as if you were about to leave it. Keep thinking of the fact that some day you will be deprived of this tenure; then you will be more brave against the necessity of departing.[53]

Yet it would be a mistake to infer from the above quotation that (a) the Stoics placed no limit on the eligible reasons for suicide; or that (b) they thought suicide was permissible without serious prior reflection.

As for (a), the only category of situations that was provisionally eligible were those which required of the individual intolerable or unworthy endurance. Ultimately, it was for the individual to decide whether his present circumstances truly met this requirement. Owing to the myriad of conditions which could sincerely be considered unendurable, a great variety of circumstances became in practice eligible for the Stoic act of suicide. Seneca, for example, argues that merely the anticipation of a considerable disturbance in one's conduct and peace of mind constituted a sufficient reason for committing suicide.[54] But he tempers this statement with the admonition that no philosopher should rush to suicide simply in order to escape ordinary forms of human suffering. Suicide is most clearly warranted, according to Seneca, only when it represents choosing ". . . an easier mode of death instead of a more painful one, thus avoiding a freak of destiny and the cruelty of man."[55] In general, vice and human folly must be met by other means than by suicide. Death furnishes no real deliverance from them since it makes the bad no better, as Edward Zeller has aptly remarked.[56]

According to one ancient authority, just five cases were normally admitted by the Stoics to warrant suicide: (1) if one's death could render a real service to others; (2) if by suiciding one could avoid being forced to commit an unlawful deed; (3) if one were impoverished, (4) chronically ill, or (5) no longer in possession of a sound mind. All of these cases are also approved by Seneca. Therefore it is fair to conclude that the Stoics *did* recognize the moral right to take one's own life. But this right was not in principle unlimited: some, not all situations, morally justified suicide. And one of these situations, as we have seen, was to be the victim of incurable or chronic disease.

In regard to (b) above, no act of suicide is considered praiseworthy by Seneca unless it is preceded by careful reflection. To Lucilius he writes:

Whichever of these ideas you ponder, you will strengthen your mind for the endurance alike of death and of life. For we need to be warned and strengthened in both directions, – not to love or to hate life overmuch; even when reason advises us to make an end of it, the impulse is not to be adopted without reflection or at headlong speed. The brave and wise man should not beat a hasty retreat from life; he should make a becoming exit. And above all, he should avoid the weakness which has taken possession of so many, – the lust for death. For just as there is an unreflecting tendency of the mind towards other things, so, my dear Lucilius, there is an unreflecting tendency towards death; this often seizes upon the noblest and most spirited men, as well as upon the craven and the abject.[57]

Furthermore, this emphasis on reflection is suggested by the favorite Stoic phrase to denote suicide, which, as we saw above, was "sensible" or "reasonable removal" (*eulogos exagoge*). This phrase implies that any defensible act of suicide must in fact manifest a rational character on the Stoic view. But, if so, what was the logic behind so-called rational forms of suicide, according to the Stoics?

Two key elements of their thinking converge to provide an answer. The first is the claim that life and death are themselves matters of indifference (morally neither good nor bad) to the wise person. The second is the claim that the reflective individual must await some evidence from destiny (God) based on an internal signal that the time to die is now. I shall discuss the second of these two claims, suggesting a serious difficulty to which it gives rise.[58]

Since life and death are for the Stoics indifferent in the sense specified above, the decision of whether or not one ought to suicide really turns on an individual's own preferences. The individual must somehow gauge whether he is living his life in accordance with nature — which was the Stoic ideal. His reason alone has the capacity to determine this, and man alone of all earthly creatures was said by the Stoics to be possessed of this divine faculty. But how could one tell whether the time for removal from life was at hand? How could one tell whether the personal conditions thought to be essential for the preservation of the Stoic way of life were irretrievably beyond reach?

By the period of the Roman Empire, later Stoics, like Seneca, began to think it was possible that the soul survives the body at death. Seneca occasionally wavered on the traditional Stoic view of death as natural personal dissolution to permit a poetic vision of divine personal immortality of the sort Plato sponsored.[59] But either under the traditional or the late Stoic concept of death, the task of the Stoic partisan reasoning about his own possible suicide remained approximately the same. Mair has plausibly characterized this task as a modified extension of Socrates' claim in the *Phaedo* that suicide was only admissible when some compelling constraint or force of circumstances (such as his own impending execution) made death unopposable. But Socrates had spoken of this compelling constraint (*ananke*) in the sense of an *external compulsion* found in one's environment. Later Stoics, including Seneca, tended to reinterpret the nature of this compulsion to mean that an *inner compulsion* need only be present. The trick was then to try to introspectively read off from one's own life situation whether any suicidal impulses one may be experiencing are in fact an omen from God that the time to leave life is at hand.[60]

This approach, of course, marks a decisive departure from Socrates' original sentiments opposing suicide in the *Phaedo*. Some would call it a sweeping rationalization for almost all varieties of suicide, reasoned or hasty, noble or base. My own opinion is that such a conceptual maneuver does indeed appear so broad as to invite conflating cases of putatively reasoned suicide, resembling the five admissible cases discussed above, with cases of merely impulsive, cowardly suicide, of the sort Seneca and his predecessors totally disowned.

Hence, the most serious flaw I have discovered in the decision procedure behind late Stoic efforts to characterize *some* forms of suicide as rational acts amounts basically to this. Any effort put forward by an individual to interpret his own life situation as one truly qualifying for suicide in accordance with the above Stoic ideals is an effort that can never be *objectively certified* as morally defensible. What's more, owing to possible self-deception, a person may misinterpret his situation and quit life for trivial reasons. A given act of suicide may, in fact, be morally defensible – at least relative to the Stoic perspective. But the final decision to commit suicide, though a limited right of every individual according to the Stoics, admits of rather few certifiably warrantable instances. Significantly, one of these instances is that of the very sick, suffering patient. But even such an unfortunate person is obliged, on the Stoic view, to weigh his plight with the greatest care, lest he quit life too soon.[61]

SUMMARY

To summarize the various positions adopted by Pythagoras, Plato, Aristotle, and Seneca on the question of suicide and euthanasia, the Pythagoreans were clearly opposed to both of these measures. Their opposition is unconditional in both cases. It was shown to rest directly on the respect for human life ethic, which is itself a moral principle derived from their logically prior religious principle that God values each embodied soul. Therefore, bodily life must not be prematurely ended, nor, as we saw earlier, prevented. At bottom, then, suicide and euthanasia are blameworthy because each embodied human soul is ordered by the gods to serve out its natural life-span: to desert one's station is tantamount to violating divine law. In a word, it is a sin, as Philolaus implies.

Plato, on the other hand, decisively sought to moderate the unconditional opposition of the Pythagoreans against suicide and euthanasia even though he shared much of their vision of death as divine personal immortality. In the *Phaedo*, Plato opposes suicide on what would have to be characterized as religious grounds. Yet even in the *Phaedo* his opposition is not unconditional. That is, if environmental constraints make imminent death irreversible, one may perhaps take early leave when no other course appears open; but that opportunity rarely occurs despite efforts by the later Stoics to broadly reinterpret this passage. In contrast, in the *Republic* Plato softens his view against suicide. This he does by approving voluntary euthanasia for the incurably ill or disabled on what amounts to utilitarian grounds. Moreover, he also approves involuntary euthanasia in the form of infanticide for defective newborns. In both instances, he links and subordinates the value of the individual to the individual's ability to perform his functions well for the state. The chronically sick and defective are seen as useless both to themselves and society. In *Laws*, Plato combines his earlier positions in the *Phaedo* and *Republic* with the added suggestion that the suicide who leaves life merely to escape his lawful duties brings shame upon himself and his community. The resulting blood pollution in this last case of cowardly self-murder is reprehensible. To demonstrate both religious offense and civil outrage an ignominious burial is required. Plato nowhere embraces the respect for life ethic as we have defined it here, else his stands favoring abortion and euthanasia would never be admitted.

As for Aristotle, he stood opposed to both suicide and euthanasia. And, it would appear, his opposition was unconditional. In the case of suicide, he thought it blameworthy not because the subject wronged himself but because he wronged his community by violating his duty to live a productive life. Hence, unlike Socrates in the *Phaedo*, Aristotle's argument is not predicated on religious considerations but rather on social considerations having to do with the proper way in which citizens should hold themselves accountable to the state. Furthermore, Aristotle offers little encouragement to proponents favoring voluntary euthanasia since he argues that death must be met bravely by the virtuous person no matter what. As the Stoics were to suggest later on (but with an altogether different twist), Aristotle felt that dying courageously — by not willfully giving into death — constituted a most significant moral test. At any rate, despite his opposition to suicide and euthanasia on moral and political grounds, it would be overhasty to characterize Aristotle as a partisan of the respect for life ethic. For he did approve what we would characterize as involuntary euthanasia for defective newborns. He also approved abortion, though this he restricted to fetuses thought to be incapable of sensation. In either case he furnished utilitarian grounds that no absolute proponent of the respect for life ethic could accept.[62]

As our principal representative of the Stoic school, Seneca, like almost all partisans of the Stoic tradition before and after him, found favor with both suicide and euthanasia. But although he held that suicide could be characterized as a rational act, and although he thought that each person had a limited right to quit life, there were at least two necessary conditions that the justified would-be suicide had to meet. First, the antecedent circumstances of the subject's proposed act must fall roughly into at least one of those five specified categories already examined which included, as we saw, being victimized by chronic or incurable disease. Second, the subject must duly reflect on his present circumstances, weighing the pros and cons in regard to himself and others. He must thereby try to insure that his proposed act does not amount to little more than an impulsive, ill-timed wish to escape life's legitimate duties, from which the virtuous person does not lightly flee through death's door. Hence Seneca's endorsement of suicide and euthanasia, though sympathetic, is not unconditional. Though he is no proponent of the so-called respect for human life ethic, that he did in the ordinary sense respect the lives of others, especially his friends, relatives, and countrymen, few would seriously deny.

CHAPTER EIGHT

THE PHYSICIAN'S MORAL RESPONSIBILITY

What sense of moral responsibility did the ancient physicians characteristically possess when granting or denying their patients' requests for abortive remedies or lethal poisons? This is the question which we are now prepared to explore. The term "responsibility" is an exceedingly rich one containing several different senses. I intend to focus here on what H. L. A. Hart has called "role responsibility." That is, I shall be inquiring into what duties or institutional constraints are borne by an individual in virtue of his or her working role in a given society.[1] My purpose here is not to reconstruct an ancient scale by which to praise or blame the conduct of the Pre-Christian physician. Rather, I aim at identifying and weighing the moral force of some of the relevant craft and related institutional values that may have influenced the typical Greek or Roman doctor's conduct when his medical assistance was sought by patients seeking to prevent or end life.

I shall argue that the dominant medical ethical values expressed by the Hippocratic Collection as a whole do not unequivocally oppose the collaboration of physicians in abortion or voluntary euthanasia. In fact, that body of authoritative medical writings furnished no unified ideological opposition to such practices. I shall further suggest that this absence of opposition by the Hippocratics and most other Pre-Christian medical sects can be partly understood as the result of a pluralistic philosophical climate in which no philosophical school succeeded in persuading the educated public that abortion and voluntary euthanasia were morally wrong. In fact, as we have seen, philosophers themselves were split on these issues: almost all but the Pythagoreans furnished some grounds for abortion, and most eventually defended euthanasia as well.

Let us turn at once to some of the more pertinent evidence from the Hippocratic Collection which fails to oppose or, in some instances, appears to allow abortion or euthanasia.

ABORTION

As for abortion, in the Hippocratic treatise *On the Development of the Child*, there exists a passage in which the author admits, in a very casual

151

and non-defensive manner, to having helped a young dancer abort her unwanted child in the earliest stages of pregnancy. The method appears crude.

... I thereupon told her to jump in such a way that the heels touched the buttocks;
she had now jumped seven times when the semen fell to the ground with a plop. And
when the woman saw this, she stared at it and exclaimed.[2]

The second century A.D. Greek gynecologist Soranus observed this apparent
contradiction regarding abortion in the Hippocratic Collection. For the Oath
had stated: "I will not give to a woman an abortive remedy." Soranus'
comments imply that in Roman times there existed at least two schools of
thought among physicians on the question of abortion. The first group,
inspired by the Oath and perhaps early Christian teaching, refused to aid in
abortion for any reason. According to Soranus, this group held that " ... the
specific task of medicine [is] to guard and preserve what has been engendered
by nature."[3] The second group, which was the larger of the two and the one
to which Soranus belonged, held that abortion was indicated whenever the
mother's life was at serious risk. Abortion for the purposes of preserving
youthful beauty or to conceal adultery, however, was explicitly condemned
by Soranus. Moreover, as we saw above, Soranus preferred contraception to
abortion because it constituted a safer remedy for the mother than destroying
her fetus, and he labored to invent many contraceptive devices to that end.[4]

Then there is the interesting testimony of the first century A.D. Roman
physician, Scribonius Largus. Scribonius was a keen admirer of Hippocrates,
whom he called "the founder of our profession."[5] Along with the first
century Roman writer Erotian, Scribonius is our earliest known source
to mention the Hippocratic Oath; happily, we possess Scribonius' comments
on some of its teachings, notably on abortion.

In the preface of his book *On Remedies*, Scribonius observes that the
Hippocratic Oath forbids abortion. He sees in this prohibition a lesson to all
in humanity. He writes: [Hippocrates] " ... has gone a long way toward
preparing the minds of learners for the love of humanity. For he who
considers it a crime to injure future life still in doubt, how much more
criminal must he then judge it to harm a full grown human being."[6] Here,
Scribonius does not explicitly state that he is himself opposed to abortion.
But these admiring remarks plus his cosmopolitan philosophy of medical
humanism, according to which "medicine promises her aid in equal measure
to all who seek her help ... and professes never to injure anyone," implicitly
sets him apart from those pagan physicians who routinely administered abortions for virtually any reason.[7] Very possibly Scribonius stands ideologically

allied with Soranus who, as we saw, criticized the unrestrained cooperation of physicians in abortion. In fact, Scribonius' preceding remarks suggest that he may have regarded the fetus as having some measure of moral worth. If so, then like the author of the Hippocratic Oath, Scribonius, too, may have been morally opposed to abortion.

It is far more certain that many ancient physicians, and I suspect most, consented to aiding women in abortion for a variety of reasons. While the Hippocratic Oath opposed even therapeutic abortions designed to spare the mother's imperiled life (since its prohition against abortion was unconditional), physicians were free to dissent, and many did so — Soranus among them.

Yet the comparatively small circle of physicians who were in complete sympathy with the Oath must have faced occasionally a serious ethical dilemma while ministering to the needs of expectant mothers. For, as Oswei Temkin has rightly emphasized, on the one hand the Oath-taker pledges to keep his patients free " . . . from harm and injustice" (P3, p. 69). On the other hand, abortion is prohibited, even for therapeutic reasons.[8] Therefore, when the prognosis indicated that the mother would be harmed or killed owing to the frailty of her health or to the expected complications of an abnormal delivery, adherents of the Oath must have faced a serious conflict of duty.[9] To be sure, the duty to obey the Oath's prohibition against abortion could clash logically (and emotionally) with the correlative duty to keep the patient from harm. Significantly, this particular dilemma found no solution from within the profession of medicine for over 2000 years. For it was not until the late nineteenth century that *safe* Caesarean section procedures and related techniques for reducing hemorrhaging and infection were invented.[10]

It is also relevant to note that the Hippocratic author of *Development of the Child*, quoted above (p. 152), apparently declined to be bound by the Oath (if, indeed, he knew of it at all). For his young dancer is a single girl, though possibly also a slave. By aborting early on she rids herself, through her doctor's cooperation, of any inconvenience that an unwanted pregnancy would have represented in her role as a musical performer. Therefore the Hippocratic physician in this instance instructs his client on how to terminate her suspected pregnancy for reasons of a non-life-threatening sort. He may well have belonged to a third group of physicians, perhaps the largest, who were willing to grant abortions for almost any reason at all.[11] Yet Soranus, who, like many of his fellow practitioners ignored the Oath on the matter of abortion, would have nonetheless found this particular girl's motives for abortion morally unacceptable. So, too, her doctor's complicity.

Hence we are again reminded of the freedom enjoyed by members of the medical craft throughout most of antiquity: physicians could ply their trade as they saw fit, subject only to the laws of the land, almost none of which barred abortion or, for that matter, voluntary euthanasia. We are also in a position to confirm that the Hippocratic teaching on abortion is by no means cut and dried, despite popular misconceptions to the contrary.

EUTHANASIA

What, then, of the Hippocratic teaching on euthanasia? Did the majority of the Hippocratic authors and their disciples advocate or oppose voluntary euthanasia? On what grounds?

The most straightforward way to approach this question is to observe that the Hippocratic authors advocated not even beginning medical treatment in cases where patients were known to be suffering from incurable illnesses. In the Hippocratic treatise *The Art*, the art of medicine (*techne iatrike*) is defined as follows: "In general terms, it is [1] to do away with the sufferings of the sick, [2] to lessen the violence of their diseases, and [3] to refuse to treat those who are over-mastered by their diseases, realizing that in such cases medicine is powerless."[12] The last part of this statement may strike some readers as evidence of a hideously stern and uncaring attitude on the part of the Greek physician. The logic behind such an attitude, however, is not far to seek. It rests on at least two motivations, one of which may be described as humanitarian, the other prudential.

First, as the author of the treatise eloquently proceeds to argue, every art or craft has its own limitations and proper ends. It is madness, he says, not to accept the fact that some diseases are by their very nature beyond the physician's power to arrest. If so, it is a dictate of reason that no thoughtful physician ought to initiate treatment when he determines in advance that a particular patient now soliciting his healing services is well beyond medical hope. To treat an incurable patient, therefore, is tantamount to willfully drawing out his suffering and raising false hopes: such a practice is unethical and smacks of bad medicine. For nature in such cases cannot be intelligently opposed.

Second, it is true that the Hippocratic physician, like his Egyptian counterpart, made extensive use of medical forecasting in order to determine rationally which patients to treat and which to leave alone. But prognostication was also useful as a prudent tool by which the doctor could simultaneously protect his own reputation by not attempting therapy on hopeless incurables whose final demise under his care might be bad for business. So this concern for reputation constituted a second motivation which was grounded primarily on economic and not simply moral or humanitarian constraints. The Hippocratic author of the treatise *Prognostic*, for example, unabashedly links the value of this art to remaining blameless in the eyes of others:

Now to restore every patient to health is impossible. To do so indeed would have been better even than forecasting the future. But as a matter of fact men do die, some owing to the severity of the disease before they summon the physician, others expiring immediately after calling him in ... It is necessary, therefore, to learn the natures of such diseases, ... and to learn how to forecast them. For in this way you will justly win respect and be an able physician. For the longer time you plan to meet each emergency, the greater your power to save those who have a change of recovery, while you will be blameless if you learn and declare beforehand those who will die and those who will get better.[13]

Furthermore, if the patient was thought to be hopeless, the Hippocratic physician was not inclined to inform him directly, unless perhaps he pleaded to be told. Instead, it was customary for the physician to inform a third party, who might in turn break the bad news as gently as possible to his sick friend or loved one.[14] Moreover, if requested to supply a lethal poison, the ancient physician was not legally barred from doing so; many apparently chose to accommodate their patient's pleas for euthanasia in just this way. Hence, the patient was not necessarily doomed to be totally abandoned by the medical community in his last days and hours of need.[15] If a physician declined to treat him, the patient could at least be made as comfortable as possible. He could also quit life altogether with the aid of fast-acting drugs sold by the willing practitioner. That such pharmaceutical assistance was often motivated by the physician's sincere compassion for his patient's unbearable suffering, I do not personally doubt.[16] If so, a decision "not to treat" cannot justifiably be ascribed merely to the ancient physician's selfish concern to protect his business reputation as a gifted healer. At least an equally important part of the story is that, on ethical grounds, many physicians sought to avoid raising false hopes or to avoid irresponsibly prolonging their patient's agony by staging costly mock treatments.

Therefore, although the Hippocratic Collection is not stocked with imperatives commanding doctors to assist consenting patients to end their tortured lives, with the notable exception of the Oath not a single Hippocratic writer voices opposition to voluntary euthanasia. Granted, this does not constitute an endorsement. But it does represent a significant fact which squares with the known medical practices of those times during which the cooperation of physicians in aiding acts of suicide among the very sick was widespread.[17] It is also a fact which further casts doubt on any popular acceptance of the Hippocratic Oath by ancient medical doctors as a whole, including those several medical writers whom we lump together under the name "Hippocrates."

So far, we have discovered that it was not generally considered morally wrong inside or outside the craft of medicine for the Hippocratic physician of the fifth and later centuries to aid in the practices of abortion or voluntary euthanasia. That is, the physician was not normally vulnerable to moral condemnation for engaging in these practices, given his vocational role as a craftsman belonging to a largely itinerant, totally unlicensed and unregulated working class which specialized in the various healing arts. In fact, we have seen that the Greek physician could practice medicine pretty much as he saw fit under the social conditions of the Hellenic and Hellenistic periods, subject only to those general legal constraints that applied to all citizens.

"DO NO HARM"

Yet in attempting to further delineate the role responsibility of the Greek and later Roman physicians in regard to abortion and euthanasia, it is pertinent to ask whether evidence from the Hippocratic Collection supports the positive claim that, if a patient so requested, it was considered by the Hippocratics and their followers to be the physician's moral duty to aid in abortion or euthanasia. Alternatively put, it is pertinent to ask whether the physician was usually liable to blame or to charges that he had violated his professional responsibilities if he declined to perform these services. It seems to me that the proper answer to these questions is no. To help support my contention, I shall re-examine one additional piece of much debated evidence from the Hippocratic texts.[18] I shall argue that this evidence, which is sometimes touted as excluding abortion or euthanasia, is in fact *neutral* on this question of there being positive moral duties to abort or to assist in voluntary euthanasia.

The evidence in question comes from *Epidemics* I. It constitutes one of the best known maxims of Greek medical literature, and it appears below, as it seldom does elsewhere, in its wider context.

Declare the past, diagnose the present, foretell the future; practice these acts. *As to diseases, make a habit of two things — to help or at least do no harm.* The art has three factors, the disease, the patient, the physician. The physician is the servant of the art. The patient must cooperate with the physician in combating the disease.[19]

Now it is not hard to imagine that in the case of a woman seeking abortion, if her physician refused assistance and she thereafter undertook to independently terminate her pregnancy by her own methods, the woman thus greatly increased her own personal risks. For the average woman, then as now, did not possess sufficient medical knowledge of her own to minimize possible damage to herself. Therefore, the doctor's refusal to abort might in some cases be fairly viewed as a contributing factor leading to a desperate patient's personal decision to take matters into her own hands, thereby very possibly doing lasting harm to herself.

These sorts of considerations make it overhasty to tag the preceding passage as one which clearly forbids physicians to help patients abort. In fact, if a particular physician did not, on philosophical grounds, conceive of the fetus as a sentient creature, nor even as a human being, then the imperative "not to harm" could thereby be understood to have no moral force in cases of abortion. And, indeed, with the exception of the implied respect for human life ethic engendered in the Hippocratic Oath, the remaining Hippocratic writings take no reasoned stand whatever on the specific question of whether the fetus is a human being whose right to exist must never be challenged by the mother, the physician, or the community.

What's more, the preceding passage deals with the physician's duties to his patient in arresting his *patient's* disease. If, therefore, the physician determined that his patient's pregnancy was threatening to harm her own life (given her present constitutional state), is it not then conceivable that the full context of this imperative could be rightly interpreted to justify abortion for therapeutic purposes? This cannot be ruled out. Even so, the passage in question admittedly does not succeed in clearly establishing any positive duty that physicians ought to perform abortions on request. At best, it remains *neutral* on that question, subject only to the partisan philosophical ideology of the attending physician.

But what about voluntary euthanasia? Don't the words " . . . at least do no harm," unequivocally condemn that practice? Probably not. For if the patient is judged to be suffering from a terminal disease, and if he is presently in the throes of great agony and is begging for death, it is far from obvious that to furnish the patient with a lethal poison in such cases is to do harm in the required sense. At least there is no reason to think it incredible that such assistance on the part of a physician might have been construed by many citizens as a kindness. This remains an open question: it would be a gross oversimplification to stereotype all Pre-Christian Greek or Roman physicians as personally favoring voluntary euthanasia. What *can* be legitimately claimed, however, is that a considerable number did. And they did so partly owing to their own partisan philosophical perspectives on the question of euthanasia — perspectives which were doubtlessly influenced to some extent by the flourishing philosophical schools of their time.

In any case, the passage under consideration from *Epidemics* I appears neutral on this issue. My analysis suggests that that passage neither entails nor excludes the ascription of positive duties on the part of the Greco-Roman physician to supply patients requesting drugs for voluntary euthanasia, or, for that matter, abortion.

SUMMARY

Nevertheless, since there is no clear ascription of such positive duties involving abortion or euthanasia throughout the Hippocratic Collection, and since the Hippocratic Oath unequivocally opposes these two practices, where does this leave us?

I have argued in Chapter Five above that the Hippocratic Oath represents a small and largely ignored reform movement within the ranks of fifth and fourth century medical practitioners. This implies that its contemporary moral force was quite limited. The Oath's prohibitions, then, were morally binding only on proportionately few of those craftsmen who practiced ancient medicine. Most physicians were not even aware of the Oath's existence. Those that were chose to adopt or ignore whatever provisions of it they felt they could stomach, based on their professional experiences, philosophical ideologies, and individual consciences. Thus, all things considered, it is likely that the following characterization comes closest to capturing the fairest picture of the average Greek doctor's moral situation in regard to abortion and voluntary euthanasia. Given his vocational role *qua* physician, the Hippocratic healer of the fifth and fourth centuries was never morally bound by positive duties to assist in abortion or voluntary euthanasia on request. He could, of course, assist in these practices without much risk of moral disapprobation from either inside or outside the ranks of his craft, if he so chose. Therefore, he possessed what may be described as a *discretionary* professional right to assist in abortion or voluntary euthanasia: he was not obliged *qua* physician either to take part or to refrain from such services. It was entirely up to his own judgment and moral sense of things.

If so, was there not also a discernible symmetry in the ancient doctor—patient relationship, especially in regard to the options of abortion and euthanasia? For, depending on the moral or religious perspectives of a given patient, is it not true that — legally, if not morally — almost all patients felt that they were similarly possessed of a parallel discretionary right to seek or decline these controversial medical services, according only as their individual consciences dictated?

This was, I submit, for a long time the case. It was not really until the Western world fell under the influence of Christianity, in the last two centuries of the Roman Empire, that the moral duties of the ancient physician became decisively redefined. When this occurred, his discretionary rights regarding abortion and euthanasia (and those also of his patients) were at first publicly condemned and then socially pressured.[20] From the early Christian perspective, the Hippocratic Oath, more than any other medical ethical work of the pagan period, contained much to be admired — though it, too, was recast to include the new deity.

For example, the earliest known Christian version of the pagan Oath, probably written some time before the third century A.D., was entitled: "From the Oath According to Hippocrates Insofar as a Christian May Swear It." Translated from the Greek, the first several lines of this Christianized version read:

Blessed be God the Father of Our Lord Jesus Christ, who is blessed for ever and ever; I lie not.

I will bring no stain upon the learning of the medical art. Neither will I give poison to anybody though asked to do so, nor will I suggest such a plan. Similarly I will not give treatment to cause abortion . . . .[21]

Surely this basic acceptance of the Hippocratic Oath was due in no small measure to the Oath's implied respect for human life ethic. For this was an ethic accorded high favor among the celebrants of this latest of the great mystery religions, just as it had won the favor of a comparatively smaller circle of religiously inspired Greek physicians many centuries earlier.

CHAPTER NINE

CONCLUSION

We have discovered not one but many diverse Greek and Roman philosophical and medical perspectives on abortion and euthanasia. I would like to glance retrospectively at some of my key findings and reflect briefly on their possible implications for a fuller understanding of the origins and character of Western medical ethics. My remarks will encompass four broad areas. In the order in which they will be addressed, these areas are: (1) the multiplicity of ancient medical ethical perspectives: (2) the relationship of ancient physicians to ancient philosophers; (3) the relationship of these physicians to their patients; and (4) the relationship of physicians to the state.

DIVERSE MEDICAL ETHICAL PERSPECTIVES

From the evidence that we have examined, manifold and diverse Greco-Roman ethical perspectives on the rightness or wrongness of abortion, euthanasia, and related topics have emerged. What does this mean?

One thing that it clearly means, and which in my opinion cannot be over-stressed, is that there is no such thing as *the* Greek or Roman view on abortion, or *the* Greek or Roman view on euthanasia, or on when humanhood begins and ends, or on what happens to a person when he dies, and so on. Some contemporary scholars, like Langer and Mair, and several more of their predecessors, like Lecky and Westermarck, have on the whole tended to lump Greco-Roman perspectives on vital bioethical and metaphysical questions like these together. Differences in degree of approval or condemnation were admitted, but overall the impression was that the Greeks and Romans of Pre-Christian antiquity generally saw eye to eye on these matters. Why? Ultimately, because of their very nearly shared values. But this tendency to construe Greco-Roman bioethical and related religious perspectives in monolithic and unitary terms is, I now suggest, mistaken.

In fact, philosophers, physicians, and laymen held a myriad of diverse views on these central questions of life and death. Some philosophers, like Pythagoras, categorically opposed abortion. Others, like Aristotle, did not. Some physicians, like the Hippocratic author of *On the Development of the Child*, performed abortions for almost any reason; Soranus, on the other

161

hand, did not. Some laymen looked forward to divine personal immortality at death; others awaited dissolution and oblivion. And on and on: one can see at a glance that not only were philosophers, physicians, and laymen free to adopt diverse ethical perspectives amongst themselves. What's more, when they did so it appears likely that their views were often formulated quite independently of one another. For example, a patient did not necessarily adopt the eschatological perspective of his physician, no matter how much the latter was revered as a capable healer. Nor, I dare say, did the philosopher easily succeed in winning his physician friends over to his moral views on infanticide, no matter how much he may otherwise have been respected for his clear-headedness. This is not to deny that there were cross-currents of influence on these and on other ethical questions. Surely there were. But the freedom to form such diverse perspectives, with little or no interference from legal or religious authorities or institutions, went a long way toward creating a ripe climate for the ethical pluralism to which this study bears witness.

Incidentally, the multiplicity of ethical perspectives so far revealed was matched, it will be recalled, by a parallel diversity of scientific approaches to the practice of medicine. Hence in the same fourth century B.C. Greek town, one could typically find physicians who would and those who would not perform abortions; one could also find physicians who were and others who were not members of a particular medical sect, and so on. Thus it is fair to conclude that Greek culture, in particular, was indeed a rich and diverse one: as much or more so in medical and in ethical matters, it turns out, as it obviously was in its ever-changing and diverse forms of artistic expression as seen in Greek architecture, vase painting, and sculpture. Moreover, doesn't this societal toleration of a multiplicity of inventive approaches to art, and to early science, and also to ethics and philosophy in general at once epitomize much of the dynamism and strength of the Greek mind?

A related question may be raised as to whether the fifth and fourth century Greeks possessed a weaker sense of moral duty than has often been ascribed to them. This question is prompted by the fact that, in medically related matters at least, a plurality of conflicting ethical perspectives flourished alongside one another. And so it is not easy to discover a single, unified paradigm expressing the ancient physician's moral duty, although the Hippocratic Oath itself has been mistakenly viewed in those terms. For there were then several different approaches to the "proper" conduct of the physician-patient relationship, though some had more currency than others.

Yet if there is an exception to this last claim, perhaps it occurs among the early Pythagorean, or Pythagorean-influenced, doctors. As I have argued, their steadfast commitment to the respect for life ethic led not only to rigorous dietary observances among the devout but also to rigorous, unitary, and unconditional medical ethical duties as well. The latter, as we saw, included the duty not to cooperate in abortion, infanticide, euthanasia, or suicide. These absolute moral constraints were formally predicated on the monolithic religious principle that God values each and every sentient creature without exception.

In addition, to a lesser extent Plato also comes close to furnishing a single moral paradigm of duty. For Plato searched for and apparently thought he had discovered the one right utopian social and moral principle from which a host of primary and secondary moral duties — many touching on matters of public health and medical care — were derivable. The principle to which I am referring, roughly speaking, expressed his view that whatever laws and customs were judged best for the smooth functioning ideal state taken as a whole were also morally right acts in terms of the particular individual's welfare. The state came first for Plato not because the individual citizen or patient did not count. Rather, it was that, all things considered, the state counted *more*: its ongoing health and very survival was seen as a necessary condition for the minimal survival of any one or more of its citizens.

Moreover, the early modern historical view that detected a decisive Pythagorean influence on Greek culture is clearly mistaken. This view was tinged by eighteenth century Neo-Platonist thinking according to which Hellenic and later Greek culture was seen as a highly synthetic whole uniquely informed by Pythagorean and Neo-Pythagorean elements.[1] Yet that early modern view, now largely abandoned, is probably not entirely wrong, as some may argue.

For, to take but one example, while Edelstein is wrong in thinking that the Oath is an exclusively Pythagorean product, he is essentially right in tracing the distinctiveness of its uncompromising respect for human life stand to Pythagorean influences of some sort. Further, I venture to suggest that in many other spheres outside of medicine and ethics, the Greeks indeed owed much to the Pythagoreans, but clearly not everything. Therefore it was truly an unfortunate exaggeration for earlier modern historians to ascribe to the Pythagoreans most of the credit for the manifold achievements of Greek culture in somewhat sweeping terms. Yet, as is sometimes the case even with historical exaggerations, there remains within them a possible glimmer of truth: the Pythagoreans, it seems, were in every sense a seminal and unique influence on Greek medicine and philosophy.

PHYSICIANS AND PHILOSOPHERS

In sizing up the relationship of ancient physicians to ancient philosophers, I have argued that in essence their feelings toward one another were both cooperative and antagonistic.

They were cooperative in the sense that both sought ultimately to plumb nature for some lasting insights into her universal laws. Many philosophers sought especailly *to explain* phenomena through these laws in order to liberate the mind from ignorance. Most physicians sought *to use* the practical consequences of these laws in order to liberate the body from suffering.

Perhaps more important for students of ethics, the intellectual relationship between physicians and philosophers was also antagonistic. For both sought to suggest right ways to live. Hence the authority of moral philosophers was implicitly challenged by doctors who — on primarily medical, not ethical grounds — directed their clients to adjust their daily regimens in accord with their favored theories of good hygiene. Conversely, the authority of doctors was effectively scrutinized by philosophers, many of whom wondered aloud (as did Plato) whether doctors were really qualified to prescribe to the ordinary citizen intimate patterns of daily living that ideally presupposed a deep understanding of the ultimate values and ends of life.

Moreover, this cooperative and antagonistic spirit has never vanished from the ranks of professional physicians and philosophers down to our own era. In the long run, this dynamic has played a constructive role in keeping each group and the public at large generally more alert to the potential problems and limitations that are engendered in these two complementary branches of humanistic and scientific endeavor.

In this connection, is it likely that part of the resourcefulness that marked the ancient physician-philosopher relationship resulted finally in each group reflecting somewhat more creatively than it had previously done on topics of life and death that had formerly belonged more exclusively to one group than to the other? That is, did some of the speculations of philosopher-scientists on such topics as when human life begins and ends, the meaning of death, the value of human life, and so on, possibly enrich the thinking of at least some of the more prominent and literate physicians on the role of their medical craft in preventing or prolonging life? Furthermore, did the widespread acquiescence of Greek and Roman physicians to their patients' requests for assistance in rendering abortions and hastening death possibly stimulate philosophers to address with greater urgency some of the moral implications of this conduct?

Personally, I do not doubt that there were cross-currents of influence in both directions. More specifically, I raised in Part Two the question of whether the views of various philosophers on the morality of abortion and euthanasia might have constituted at least one contributing cause to the eventual social acceptance of these practices in antiquity. I am inclined to answer now in the affirmative. Admittedly, the evidence I have marshalled in the present study must in fairness be called inconclusive on this question. I have furnished a detailed account of mainly four prominent philosophers on these topics, and the writings of many more remain forever irreclaimable in the ashes of history.

But of the four philosophers herein examined, it is worth noting that three argued in favor of abortion and infanticide under certain conditions (Plato, Aristotle, and Seneca). In addition, two endorsed the moral acceptability of voluntary euthanasia for the terminally or chronically ill (Plato, Seneca). These results surely add strength to the speculation that Greco-Roman philosophers played some role in fostering the dominant moral attitudes of a good many doctors and patients who felt that abortion and euthanasia were morally defensible. But I hesitate to say that this has been proved.

First, there is a formidable causal objection to contend with, one that the British philosopher David Hume so clearly defined.[2] That is, even if there existed sufficient evidence to show that most leading Greek or Roman philosophers judged abortion and euthanasia to be morally acceptable, one could not logically assert on this prior evidence alone that philosophers therefore succeeded in influencing the masses that these two practices were all right. Much too much is taken for granted in such a hasty judgment: for (1) that B constantly follows A does not entail that A necessarily causes B; and (2) there might be a variety of contributing causes for B, many of which we fail to perceive or distinguish.

Second, the possibility of a reverse assimilation of values cannot be fairly dismissed as a viable objection either. It is undeniable that on occasion the popular morals of the masses have in subtle ways infected the moral perspectives of academically trained philosophers or other leading experts.[3] I dare say this sort of thing goes on in our own epoch as it did in antiquity; it might be described as a trickle-up theory of social norms and values. Though I cannot stop to debate the possible merits of such an interesting sociological theory here, these preceding objections jointly constitute a real challenge to anyone attempting to gauge the historic influence of moral philosophers in shaping the medical or ethical values of a given culture — past, present, or future. Though I do not regard these objections as necessarily fatal, they nonetheless give serious pause to any thoughtful custodian of the history of ideas.

PHYSICIANS AND PATIENTS

There are two additional implications that come immediately to mind in view of my analysis of the moral responsibility of the physician toward his patients in responding to their requests for abortion or euthanasia. These are that the typical Greek or Roman doctor felt that he had neither (1) the absolute duty to prolong human life nor (2) the absolute duty to protect such life at any cost. I would like to speculate a bit on why this may have been so.

As for the absolute duty to *prolong* human life, I have already argued that the Hippocratic maxim " ... First, do no harm," was really neutral on the question of whether a physician should assist in voluntary euthanasia. Furthermore, the average Greco-Roman physician was trained to assist the forces of nature in finding a natural resolution of patient discomfort and disease. This attitude followed in part from what I earlier referred to in Chapter Two as the Hippocratic Principle of Natural Healing. This principle recognized that the body possessed its own biological rhythms and curative powers. The physician's job was to discover how these forces worked and then do nothing that would impede nature in her normal, curative course.

What is often missed, however, is that part of the Hippocratic physician's sense of the incalculable powers of nature as an indwelling force affecting the constitution of each and every patient rested on his accompanying awareness that death, too, represented a natural resolution to the human suffering generated by certain forms of disease. Of course, death was never greeted by the physician as a welcome visitor whenever his creative diagnostic and therapeutic skills proved equal to the challenge of a particular illness. But at other less fruitful times, to oppose nature in her normal cycle of generation and decay seemed the equivalent of hubris — especially when no known medical remedy appeared to exist.

To these considerations, also, I venture the following account. As the popular horror of death and the prospect of dwelling in a dank and unfriendly Hades eventually subsided, due in part to Aristotelian, Epicurean, and Stoic efforts to debunk the terrors of the afterworld, the very thought of assisting a patient in suicide or of declining heroic measures to prolong the life of a suffering victim may have seemed somewhat less objectionable. For aside from the fact that death could then be more plausibly conceived by physician and patient alike as a refuge from untreatable pain and despair, there was also less cause to worry on religious grounds that the gods would thereby be offended. Or to worry that, as the earlier Pythagoreans had claimed, the suicide's soul would suffer a double penance in the next life. In a word, the prospect of death was just not that scary any more.

As for the absolute duty to *protect* human life at all times, not only from diseases but from various members of the community who seek to prevent or end life (as in requests for abortion), the typical Greek or Roman physician felt unbound by such absolute constraints. In the case of requests for abortion, one cannot help but wonder whether part of the willingness of many physicians to grant abortions had something to do with their individual views on the status of the fetus within the moral community. Should the fetus be regarded as a person protected by the minimal moral and legal protections enjoyed by other children or by adults?

As we saw, the popular answer to this question by most laymen was no. In addition, except for the Pythagoreans, none of the philosophers that we examined held that human life began at the moment of conception in the relevant sense required here. Even though I did not find anywhere in the Hippocratic Collection an explicit position taken on this controversial issue by any one of the authors, it is very possible that the popular estimate of the fetus's low status may have carried over into medical appraisals of this matter, too. Also, there is the distinct possibility that various philosophical schools may have affected the thinking of physicians on this question as well, though which schools were most favored is hard to say.

It is with greater confidence that one can say which schools were in disfavor. Here it is truly ironic that the Pythagorean school comes readily to mind. For until rather recently most scholars and the general public had associated the Hippocratic Oath and its respect for human life ethic with the dominant ethical philosophy of the ancient medical community.

Yet although this stereotype has now been cast into doubt, one must beware of exaggerating the picture of the ancient physician-patient relationship in the opposite extreme. I, for one, think that a fair number of Greco-Roman physicians may well have recognized at least a presumptive or *prima facie* duty to protect human life in something like W. D. Ross' sense, though not in any absolute sense. That is, all things being equal, they would attempt to protect human life, though possibly to allow overriding reasons on a case-by-case basis for not doing so.[4] True, the average Greek physician probably did not swear to uphold the famous Oath of Hippocrates. He may never have even heard of it. But as a trained healer, wasn't it likely that his initial moral bias tended to favor the protection rather than destruction of human life?

In the special case of abortion, the moral issue additionally turned on whether the fetus possessed an incontestable right to be born. That such a right was routinely denied not only in philosophical circles, but in religious and legal circles as well, fostered a climate of institutional approval or inattention to abortion throughout most of antiquity. At bottom, the respect for human life ethic found few supporters. As I suggested at the close of the preceding chapter, it expressed a moral value that was all but ignored by most pagan physicians even though they themselves may have been among its first authors. It remained for the early Christians to insist on that ethic's divine moral status and declare it the absolute standard of right conduct for doctor and patient alike.

PHYSICIANS AND THE STATE

We have seen that the Greeks were not the first ancient people to attempt to define the proper conduct of their physicians. The Babylonian Code of Hammurabi antedates the rise of Greek medical ethics by well over 1,000 years. Moreover, the Assyrians, Persians, and Egyptians were shown to have made some efforts to formalize the duties of their physician-priests in regard to abortion or euthanasia.

Yet there was a marked difference in the Greek social and political conditions which accompanied the efforts of at least some fifth century Hellenic physicians to begin inventing and adopting moral guidelines for the proper practice of medicine. That difference was that the Greek city-states by and large left their physicians alone. Greek physicians were virtually free to accept or reject patients and treat them in accordance with their own best judgment. The Greek states did not impose craft restrictions and additional legal regulations on their practitioners, unlike the states of Assyro-Babylonia, Persia, and Egypt, where the rulers busied themselves with such domestic affairs.

Hence it is fair to ask what conditions may have been responsible for the doctor-initiated flowering of Greek medical ethics and etiquette? What would have caused at least some physicians to voluntarily invent or adopt various medical Oaths or rules of craft conduct in the first place?

One possible answer is to attempt to explain this achievement as the fortuitous outcome of the so-called explosion of Greek genius which is popularly associated with the philosophic, scientific, political, and artistic advances of the fifth century Athenian Golden Age. But it strikes me that the Greek love of abstract ideas and the cultural momentum of the Golden Age, however real these influences may have been on the general advance of

Greek scientific medicine, supply all-too-ready answers to our query here, and answers that finally explain very little.

More plausibly, I think the physician-initiated and voluntarily imposed Hippocratic norms can be explained as the partial result of a felt need on the part of some practitioners to bring more uniformity and integrity to the conduct of physicians as a whole — both toward each other, and toward their patients.

Thus we observed of the Hippocratic Oath that is covenant specifies some of the more important duties between established master-physicians and their junior partners whom they undertook to properly train. Similarly, the four Hippocratic works on etiquette also contained maxims which aimed at regularizing how physicians should act toward one another during professional consultations and in the marketplace. Such physician concerns can, of course, be properly understood from the altruistic standpoint of wishing to protect the patient's best interests; by sparing him, for example, from unpleasant bedside disputations between two doctors quarreling over the patient's proper treatment. Or, typically, by sparing him from the abuses of would-be doctors or charlatans who had not been properly schooled in the trade secrets of the healing art.

Yet that is only part of the story. These identical medical concerns must also be frankly recognized for their egoistic content. They held the potential to protect the professional and business reputations of the ancient physicians themselves, who, after all, sought to sell their services to a trusting and respectful public.

Therefore it is fair to conclude that, at bottom, the primary elements contributing to the emergence of Greco-Roman medical ethics and etiquette were twofold. First, there was the pre-existing condition of freedom. Second, there was the willingness of some Greek physicians to respond to this freedom by accepting responsibility for their own professional conduct and that of their entire craft. Freedom brought with it the opportunity for medical experimentation and the chance for progressive advances in patient care. The very same freedom, however, also fostered opportunities for charlatanism, medical chicanery, and higher than acceptable risks of abuse or permanent injury to patients and their families who were seeking medical services. This latter prospect, in particular, struck a good many Greek and Roman physicians as morally unacceptable. It also left their craft open to hostile forms of public suspicion and criticism, some justified and some slanderous. This criticism and suspicion was bad for the business of ancient medicine; it represents a professional vulnerability from which medicine has not yet entirely escaped to this day.

CHAPTER TEN

EPILOGUE

FUTURE PROSPECTS

It is tempting to call the present investigation of Greek and Roman medical ethics to a halt on the grounds that nothing else can be informatively said. But in fact, these and related ancient topics of contemporary import are deserving of further study and comment in a number of respects. I shall briefly indicate what a few of these studies might entail.

To begin with, there are two areas of ethical and legal interest that have not here been fully addressed: (1) medical experimentation on patients and (2) maintaining the confidentiality of the patient's health, prognosis, and treatment. Some work in the legal area has been undertaken by Amundsen. His studies suggest that, contrary to traditional scholarly opinion, physicians were in some cases subject to malpractice suits in both Greece and Rome.[1] Yet the specific question remains: were intentional breeches of patient confidentiality covered by legal penalties of any kind? Probably not, but there may still be room for investigation concerning this matter. One research area that promises to be fruitful is the rhetorical literature. Mock legal defenses were created by debating contestants in order to artfully demonstrate their oratorical skills. This they did through the evaluation of various political and moral issues, a few of which touched on the physician's role in society and so might tell us even more.

Edelstein has explored part of the question of the medical experimentation on humans during ancient times in an interesting paper devoted primarily to the history of anatomy.[2] But as beneficial as his general account has proved, a more specialized estimate of just how extensive the practice of vivisection was among slaves or condemned prisoners throughout Greek and Roman antiquity would be most welcome. Gary Ferngren has moved a step in this direction with a recent article centering on vivisection during the Roman period.[3] Future studies might uncover additional details about the moral reasoning and motives of those physicians who engaged in this and related practices. An even broader understanding of the value of human life at different times in Greek and Roman history may thereby be secured.

In addition, a systematic examination of Greek and Roman attitudes toward the mentally ill would be profitable. This could cast more light on their moral views concerning the possible blameworthiness of those who committed suicide impulsively – in the absence of the sort of "sensible reflection" which the Stoics and others thought necessary to dignify (if not morally justify) the act. George Rosen has done some preliminary work in this area.[4] But his focus has been primarily sociological and historical, not ethical. Moreover, E. R. Dodd's provocative book, *The Greeks and the Irrational*, broadly explores how the Greeks viewed the emotions in connection with self-destructive behavior. Yet his overall thesis extends far beyond ethical topics, which, for the most part, are set aside. Even so, Rosen's and Dodd's research furnishes a useful starting point for a fuller analysis of the moral appraisal of euthanasia and suicide in antiquity for those special cases where mental illness is involved.

Finally, in further seeking to explore the origins, influences, and significance of the Hippocratic Oath and ancient medical ethics generally, at least three additional avenues of research come to mind. Of these, one in particular appears especially worthwhile.

The first would involve attempting to go still farther back in time in order to penetrate the earliest beginnings of Greek medical values and practices during the so-called Greek Dark Ages. This refers to a time roughly between the ninth and sixth centuries B.C. But since we possess almost no written records at all of Greek medical activities from the death of Homer to the birth of Hippocrates, such an attempt must be indefinitely postponed. Regrettably, the requisite evidence is lacking.

The second avenue would involve resuming the present account where I have left off. This would entail exploring the writings of the early Church fathers and related authors of the late Roman and early Medieval periods in order to evaluate further the impact of Hippocratic idealism on subsequent European medical values up to the period of the Renaissance. To date, Loren MacKinney has sifted through virtually all of the known early Christian medical documents bearing on the medical ethics and etiquette of most of the period in question.[5] Also, in a somewhat earlier article, Pearl Kibre has profitably focused on "Hippocratic Writings in the Middle Ages." Thus, much of this proposed line of inquiry has already been accomplished. At present, it is not at all clear whether anything really new would surface by merely re-examining the evidence which these two writers have so capably analyzed. But perhaps more could be learned of medical ethical developments during the Renaissance in Italy and elsewhere.

The third avenue is the most promising of the three. It would involve a detailed *comparative* study of contemporary Western ethical perspectives on abortion, euthanasia, and related issues with the earlier ancient Greek and Roman perspectives which this present inquiry has explored. Much could be gained from such a comparative study: especially a firmer sense of whether the basic ethical principles which undergird many of our most popular contemporary arguments concerning these issues were also principles that were known and used by the ancients in one form or other. My guess is that in many instances the essential arguments for and against abortion or euthanasia would turn out to be very similar. For example, many features of Plato's eugenic argument in favor of infanticide for defective newborns, plus some of the Stoic arguments supporting abortion, are already being revived or rephrased in contemporary philosophic, journalistic, and legislative circles. If so, does this imply that there exists a single common fund of such arguments, both pro and con? That remains an open question. Yet given the creative human potential for generating fresh ethical appraisals and the unique moral problems to which modern medical technology has given rise, one must surely doubt that contemporary medical ethical thinking has remained a mere echo of past philosophic traditions.

In any case, this long glance backward into Greco-Roman times may be regarded as something more than a self-indulgent excursion if, as I hope, it ultimately serves to enrich public and scholarly debate on these and other pressing moral matters. I believe that such enrichment can occur in at least two ways.

First, an awareness of the ancient roots of Western medical values and conduct can help us locate and clarify the more enduring elements of our contemporary ethical controversies. The present study has in part aimed at furthering just this goal. Moreover, this renewed public awareness can assist us in separating popular myths about how abortion, infanticide, euthanasia, and suicide have been regarded in our classical past from the actual facts. Such debunking is not only necessary but liberating. For it appears that misleading arguments by authority based on alleged ancient medical tradition still find occasional champions in current discussions of these and other bioethical issues. If so, the more informed we are about our moral and medical heritage, the less likely we are to be tricked by such oversimplified, misleading reasoning. Such reasoning typically fails to take into account the plurality of ethical views held by physicians throughout antiquity before and well after the birth of Christianity. So, too, the more demanding of historical precision we become, the more likely we are to win for ourselves and others a finer understanding of the dynamic cultural heritage and values that continue to shape our choices and conduct even today in the subtlest of ways.

Second, by determining whether many of the same — or essentially different — issues and arguments constituted the fulcrum of bioethical debate for the ancients, the contemporary student of ethics is left in the most favorable practical position to accept, modify, reject, or invent fresh appraisals and alternative solutions to these perennial life and death problems. Whatever conclusions are finally reached, much can be gained, including a deeper appreciation and sense of kinship with our Greek and Roman fore-bears, as well as a more refined sensitivity to the unique moral problems faced by twentieth century medicine.

A CONTEMPORARY APPRAISAL

The Biomedical Revolution

As much as we may admire the therapeutic and ethical achievements of the Hippocratic physicians and their successors in Greece and Rome, most agree that the physician-patient relationship has undergone many significant changes since antiquity. Much of this change is due to the comparatively recent invention and application of automated medical technologies and experimental drug therapies in the treatment of disease. For example, in ever more instances physicians can keep brain-dead patients "alive" almost indefinitely through the use of the respirator machine which sustains their breathing and aids circulation. These are patients many of whom would have died of their afflictions straightaway in earlier, pre-technological times. Moreover, this ethically perplexing state of affairs in medicine has been realized by a vast array of sophisticated critical care appliances, as well as fully coordinated monitors, powerful chemotherapies, new surgical tech-niques, and, most recently, mechanical and donor organ substitutes.

In stark contrast, the ancient physician did not face the awesome problem of being able to prolong a patient's life indefinitely. He lived during a com-paratively pre-technological period when the latest advances in medical technology included new forms of bandaging, drug and dietary measures, pain relieving surgical maneuvers, or the application of palpating techniques. He literally had no occasion, therefore, to agonize over whether his terminally ill kidney patient would be ready to receive a compatible donor organ in time to avert impending renal failure. Nor, obviously, did he have any care about many of the brand-new ethical problems presented today by such controversial procedures as the formation of human life in the test tube (*in vitro* fertilization). Nor was he faced with the counseling dilemmas brought on by genetic screening. These include deciding whether to share or withhold sensitive genetic information to prospective parents regarding the health of their yet unborn offspring, based on what is often merely statistical, inconclusive evidence. Nor, again, was the ancient physician bedeviled to the same degree as his modern counterpart by the option to prescribe an increasingly complex series of largely untested experimental drugs to consenting, terminally ill patients pleading for relief or cure at any cost. These are but a few of the many pressing ethical dilemmas which now confront contemporary teams of medical specialists trying to protect and save lives in medical centers across the country. Most if not all of these dilemmas involve fresh medical options and treatment strategies spawned by the biological, engineering, and medical revolutions which have converged in the twentieth century. As a result, nothing quite so exciting or unsettling has occurred before either in the history of medicine or in the evolution of medical ethics.

Given this unique technological character of contemporary medicine, some critics have objected that the Hippocratic medical tradition − its patient care concepts and ethical principles in particular − is manifestly irrelevant. For the Hippocratic tradition came into being in comparatively primitive times vastly different from our own. Such critics therefore conclude that the moral rules associated with the Hippocratic legacy are entirely inappropriate for contemporary ethical problem-solving. The sooner this is recognized, they say, the better. Specifically, some object to the strictures against abortion and voluntary euthanasia embodied in the Hippocratic Oath since they are phrased in absolute terms, admitting no exceptions whatever. These two prohibitions, along with restrictions against infanticide and suicide which the Oath tacitly implies, need to be loosened or abandoned, critics allege, in view of biomedical developments like amniocentesis and heart-lung machines which have greatly expanded our options.

Consequently, many thoughtful people both inside and outside of medicine now question whether all cases of abortion or voluntary euthanasia are morally wrong, as the Oath declares. What's more, landmark United States Supreme Court decisions like *Roe vs. Wade* (1973), which, among other things, legalized abortion on request for virtually any reason during the first trimester of a woman's pregnancy, unmistakably signal society's growing skepticism toward the correctness of certain Hippocratic ideals. Very likely, the highest Court or the Federal legislature will similarly be impelled to reconsider the question of *active* voluntary euthanasia — which is now strictly illegal — if the present mood of the public persists. As it stands, a score of states are presently considering passage of legislation similar to the Natural Death Act which was passed into law in California in 1976. This kind of legislation grants patients the right to inform physicians by a statement signed in advance that no extraordinary measures should be taken to prolong their lives should they ever have an incurable injury, disease, or illness that is certifiably terminal. But since, as we have seen, the Hippocratic doctors and their heirs were equally disinclined to prolong the lives of hopelessly suffering patients, it is wrong to infer that such legislation violates the spirit or letter of the Hippocratic tradition *per se*.

In view of the controversial nature of these and other bioethical issues, it is hardly surprising to find that the American Medical Association remains silent on the moral justifiability of abortion and euthanasia in its most recent (1980) Code of Ethics. (See Appendix A, p. 188). Surely it is fair to conclude that the ethics of medicine now, as in antiquity, remains entirely vulnerable to the influences of popular and religious moral sentiment. Twentieth-century physicians, like the patients they serve, hold a myriad of diverse personal views on these topics. They are not, like the rare ancient Oath-taker who swore to Apollo, of one and only one mind. Thus, a great measure of latitude is left to the modern physician's individual conscience. In this sense, doctors today have much in common with the majority of their ancient Greek and Roman brethren, most of whom were similarly repelled by the single-mindedness of the Hippocratic Oath (if they ever heard of it at all), and so paid it little or no heed.

Even more, some critics object to what they describe as the overbearing paternalism toward patients which the Hippocratic writings evince. The ancient idea that, all things considered, the physician knows what is best for his patient's welfare is an assumption increasingly at odds with the general public. So much so that the burgeoning patient rights movement in America and elsewhere has sought to render that attitude obsolete. This movement asserts that normally it is the patient-consumer's right to know the full truth about his diagnosis, treatment, and prognosis in plain language which the layman can be reasonably expected to understand. (See Appendix B, p. 189.) It also emphasizes the right of the patient to be frankly told of viable alternatives to the therapy his or her physician is prescribing.[6] This serves the purpose, among other alleged benefits, of involving the patient more directly and honestly in the selection of the most cost-effective and therapeutically helpful treatment. Yet all the uproar over patient rights has occurred against the backdrop of a seemingly antiquarian Hippocratic value system, one in which the physician unabashedly claims to possess superior judgment plus a thoroughgoing paternalistic attitude which is beyond reproach.

For example, in the Hippocratic Oath it is proclaimed: "I will apply dietetic measures for the benefit of the sick according to my ability and judgment" (see above, P3, p. 69). In *Decorum* the physician is advised to "reveal nothing of the patient's future or present condition, for many patients on account of this have taken a turn for the worse."[7] This last bit of advice no doubt applied mainly to patients with the more serious afflictions. However, the paternalistic voice of Hippocratic medicine – now often spurned – is unmistakable. In ancient Rome, Galen later went so far as to suggest that the patient should be encouraged to admire his physician like a god, since this reverential attitude would supposedly speed the healing process.

Furthermore, Hippocratic medicine has been faulted for its narrow tendency to define medical care mainly in terms of emergency treatment.[8] It allegedly overlooks the allied function of medicine as a body of knowledge and battery of services equally useful to the maintenance of health and the prevention of disease. On this score, it is undeniably true that in the decade of the 1970's so-called "preventive medicine" has received renewed emphasis through such channels as public information advertising, voluntary genetic screening programs, and a general emphasis placed by health professionals on sound personal regimen. But as to the latter, it is simply mistaken to construe the ancient physicians as healers who disregarded the importance of keeping their patients well through proper diet, exercise, rest, and hygiene. The patient was encouraged whenever possible to accept responsibility for the balance and maintenance of his bodily constitution. Detailed accounts of sound regimen were in fact regularly supplied in the medical literature. Typical of this preventive health orientation is the advice of the fourth century B.C. physician, Diocles of Carustus. He writes: "The cultivation of health begins with the moment a man wakes up."[9] In fact, it was usually within the conceptual milieu of a holistic approach to health care, one which stressed the importance of both mental and physical factors in regaining and preserving well-being, that the ancient physician and his patient worked. That modern medicine has lately turned with new appreciation to the concepts of holistic health and has also rediscovered the value of preventive medicine, serves to remind us of the amazing resiliency and relevance of the Hippocratic medical tradition just here.

Moreover, in contrast to the ancient Greek or Roman physician who typically worked alone or with a student helper, the modern attending physician functions in the hospital setting as one member of a complex team of primary and secondary care-givers. Granted, the attending physician is customarily in charge of his case. But he may arrive at his diagnosis and implement his treatment plan in collaboration with medical specialists from a wide array of technical backgrounds. Hence the specialist in internal medicine faced with a difficult case might draw upon the expertise of a hematologist, radiologist, ophthalmologist, dermatologist, and neurologist before settling on a final diagnosis. Registered nurses, physical therapists, lab technicians, paramedics, and social workers may also contribute vital services aimed at helping the patient recover.

While all this is so, it is equally true that seeking diagnostic advice from other medically trained practitioners was never really opposed by the Hippocratic authors. Just the opposite: it was encouraged. Unlike charlatans, good physicians are described by these writers as being able to recognize their own limitations and harboring respect for the sometimes superior knowledge of their colleagues without feelings of jealousy. For example, the Hippocratic author of *Precepts* emphatically states: "A physician does not violate etiquette even if, being in difficulties on occasion over a patient and uncertain owing to inexperience, he should urge the calling in of others in order to learn by consultation the truth about the case."[10] Of course, it was acknowledged that medical consultations could produce irritating consequences on occasion, just as they may today. So, with an eye to proper decorum and the desire not to alarm the patient uppermost in his mind, that same author prudently admonishes the consulting doctors to " . . . never quarrel or jeer at one another."

Yet despite the openness of the Hippocratic physician to diagnostic assistance from peers, it is true that he was in general distrustful of medical assistance being rendered from mere family members, well-intentioned laymen, and others *outside* the medical fraternity. For one thing, laymen were said to be bad at following orders; for another, it was pointed out that they were usually the first to blame the doctor if the patient took an unfortunate turn for the worse.[11] Also, it is evident that specialized nurses (midwives excepted), physical therapists, lab technicians, social workers, and related support personal stand without any real parallel in the ancient setting. But this hardly seems a serious indictment of Hippocratic medicine when taken in context. Rather, it points to the frankly simpler administration and understanding of the healing arts at that time. Given the strictly laissez-faire, unregulated, self-sufficient fashion in which medicine was then practiced, it would have been entirely unnatural, I contend, to involve anything like an orchestrated medical team in the treatment of acute injury or illness. For, at bottom, what would have been gained — therapeutically, financially, or otherwise — in a vocational framework still largely bereft of specialization and institutionalization?

Nevertheless, in view of the preceding catalogue of modern medical trends toward extraordinary high technologies, the patient rights movement, preventive health measures, and hospital team care, it is hard to resist the conclusion that contemporary medicine differs in many striking ways from that antique brand of healing known to Homer, Hippocrates, Aristotle, and Galen. But do all these changes amount to a difference in the degree or in the kind of medicine that is being parcticed today? If the kind of medicine being practiced today is radically different, should we not then resolutely, and once and for all, set aside the Hippocratic Oath (let alone the larger Hippocratic writings with which it is loosely associated) as " . . . a museum piece, an amiable but useless custom"?

These are the provocative words of the physician and classicist Carleton B. Chapman. In a recent *New York Times* feature story focusing on the Hippocratic Oath's role in American medical school education, Chapman is quoted as saying: "For centuries the Oath has been viewed sentimentally and uncritically by the medical profession as its ethical standard and has, rather paradoxically, served as a barrier to the development of an adequate and comprehensible ethical statement for the profession."[12] Also, in the United States few, if any, graduating classes of medical students are required to recite the Oath anymore. And almost none is required to study it. So, what, if anything, can be said in favor of the Hippocratic Oath or the Hippocratic writings on etiquette in view of such alleged deficiencies?

That the Oath and the name of Hippocrates may still remain in the popular mind the highest symbol of the physician's dedication to healing, or that it may continue to encourage a fundamental bond of trust between the patient, the physician, and the medical profession, matters not. For if this popular impression is predicated on a romantic vision of Hippocratic medicine which — we now know — was at best rarely adopted in antiquity, and which, in addition, has now been largely replaced by more serviceable medical codes in our own epoch, then there is little consolation to be derived from such misplaced popular enthusiasm. Ultimately, such enthusiasm is based not only on an ignorance of the past but on a misunderstanding of the real moral and medical needs of the present and future. In fact, it is our duty to set aside the Hippocratic Oath entirely, I would insist, if it can be plausibly demonstrated that its moral and humanitarian ideals have substantially outlived their usefulness.

But has this happened? Far from it. In fact, I shall argue that the Oath, and the reform movement in medicine from which it partly derived, is still surprisingly relevant to contemporary medical ethical concerns. This is not to say that the Oath is flawless. It is not to say that one could never be completely justified in overruling its rigidly conceived duties. I readily grant that many significant changes have been introduced into the physician-patient relationship during the biomedical revolution of this century. But I further contend that the net effect represents an overall change in the degree rather than in the kind of basic moral awareness that is still needed on the part of health professionals in order to insure that the most humane medical services possible will continue to be provided in the decades to come.

In my opinion, the Oath remains surprisingly relevant in at least two respects. First, it continues to center the attention of society on a basic core of perennial medical ethical issues that constitute virtually ineliminable categories of legitimate moral concern that we ignore only at our own peril.

Second, it affirms the importance of the graduate physician's covenant (or public promise) to be professionally committed to and so morally responsible for: (a) the physical and psychological welfare of patients who trustingly, and often naively, seek medical relief from their assorted maladies; and (b) the technical competence and moral integrity of himself as well as his or her medically trained peers.

In the remaining pages of this chapter, I would like to explore briefly these two contemporary contributions which the Oath continues to make.

PERENNIAL ISSUES

As for my first claim, the Hippocratic Oath may certainly be credited with usefully focusing the attention of Western medicine on at least five central medical ethical issues. These core issues are: (1) patient confidentiality; (2) abortion; (3) euthanasia; (4) truth-telling; and (5) justice in the distribution of health services. In my opinion, this list serves to highlight, though it far from exhausts, crucial areas of personal and institutional conduct that no comprehensive medical code for physicians or other health professionals can legitimately ignore. To do so would amount to a crass desertion of among the most fundamental medical ethical concerns of our own, or any other, humane era.

Incidentally, if the Oath's stricture against surgery strikes the modern reader as unforgivably foolish (see P5, p. 69), as indeed it may, given the many fine surgical achievements of the present century, this defect is to my mind decisively outweighed by the more progressive features of the Oath which remain substantially pertinent. It should be accepted for what it is: an anachronism. Even so, the reservations of the Oath against surgery are at least understandable when one recalls the general crudeness of surgical techniques in antiquity, many applications of which killed more patients than were cured. (Recall, too, that the Oath never forbade cautery or bone-setting explicitly). For that matter, one cannot help noticing that the issue of unnecessary surgery is still being debated on several fronts by surgeons and their qualified critics in our own time and place.

In any case, the ethical topics (1)–(5) above addressed by the Oath and supplemented in the related writings on etiquette deserve a fair hearing; let us consider some aspects of them now in the fuller context of twentieth century developments.

Confidentiality

The Oath implies an intelligent concern for the confidentiality of the patient's prior medical history, prescribed treatment, and personal life. It states, in part: "Whatever I may see or hear in the course of treatment . . . I will keep to myself, holding such things shameful to be spoken about" (see P7, p. 69, above). However, some critics fault the Oath on this. They take the utilitarian position that *sometimes* a greater good would accrue to a greater number of people in society if physicians were willing to inform law enforcement officials about any suspicious details of a given patient's mental or physical health. For example, couldn't some crimes be prevented if psychiatrists were legally required to alert local authorities about any patients in their care reportedly having homicidal dreams? And what about patients who arrive at emergency rooms bearing unexplained cuts or wounds?

Proponents of such overriding exceptions to the confidentiality rule face the burden of devising a sound referral criterion that would insure that no patients could be singled out arbitrarily. Obviously, almost any such criterion could lead to unintended abuses. There is always the possibility of misjudging the true nature of the patient's present condition (or his past or future conduct). Thus, it is fair to conclude that, at very least, the presumptive duty of the physician to preserve the privacy of his patient's affairs is admirably defended by the Oath. If exceptions are to be granted, and conceivably this may be necessary to do on rare occasions, every precaution must be adopted to promote even-handedness and a sound rationale for doing so.

Abortion and Euthanasia

The Hippocratic Oath also relevantly centers our attention on the value and respect we may or may not ascribe to human life. This it does by pronouncing its twin strictures against abortion and euthanasia in uncompromising language. As individuals, we may well challenge the absolutist position that the Oath takes in opposing these acts. For example, some may plausibly argue that when the mother's physical or psychological condition is seriously threatened by the fetus, abortion is then morally justified. Or, in reference to euthanasia, some may argue that when a terminally ill patient is suffering torturous pain of such magnitude that (a) no drug will any longer furnish relief, and (b) the patient pleads to have his life ended, active voluntary euthanasia via a lethal injection is then morally justified. Whether these or numerous other eligible dissenting positions on abortion and euthanasia turn out to make more moral sense than the Oath's position will depend ultimately, it seems to me, on the specifics of each case and the soundness of the reasons advanced pro or con. But the simple point here is that, minimally, by condemning these acts the Hippocratic Oath succeeds in dramatically calling our attention to two crucial matters of moral choice. These matters are no less important for us to clarify and deal with today than they were in the Athens of Socrates.

So here, too, I conclude that the Oath provisionally serves as a reliable introductory guide to moral considerations that are inextricably linked to the most momentous medical dilemmas, namely, judgments affecting who shall live and who shall die. At bottom, the Oath proclaims a bias favoring life and the living for which no apology need be made.

Truth-Telling

Regarding the physician's duty to be truthful, the Oath-taker pledges: "I will come for the benefit of the sick, remaining free of all intentional unjustice, of all mischief . . . " (see P6, p. 69 above). This paternalistic principle implies that the doctor ought to do whatever he thinks will benefit the sick person in his care. Also, he should not lie, for this would be a form of mischief. He should carry out his course of treatment in a forthright, physician-knows-best manner " . . . according to my ability and judgment," as the Oath says (P3).

Yet elsewhere in the Hippocratic writings on etiquette, as we saw earlier at *Decorum* XVI, the physician is urged not to reveal to his patients any gloomy assessment of their illnesses without careful forethought about the possible demoralizing effects it could have on them. The result is that on the one hand there is the moral duty not to lie; this the entire Oath, being a public promise, presupposes. On the other hand, this very duty sometimes comes into conflict with the moral rule not to knowingly harm the patient (which P3 of the Oath also underscores). Regrettably, the Oath neither acknowledges nor resolves this problem of a possible conflict of duty. Being the kind of compactly worded religious or guild document that it is, one could not fairly expect it to address every conceivable difficulty to which it might in practice give rise. But I suspect that the rule against intentionally harming the patient was routinely allowed to override the competing rule against deception in those instances where the brutal truth was judged by the more caring doctors to be much too shattering for a particular patient to bear.

Significantly, how best to settle such delicate circumstances as these is still very much part of a contemporary debate that continues to rage in medical quarters. For example, the authors of the much heralded "A Patient's Bill of Rights," issued in 1973 by the American Hospital Association, also appear willing to yield to the paternalistic principle of not harming the patient when this latter principle conflicts with the patient's alleged right to know his or her own true medical condition (see Appendix B, p. 189). But exactly when ought the truth-telling rule be overridden? Presumably when directly telling the patient the truth "is not medically advisable," as the authors of the Bill have vaguely written – thereby leaving the matter to the physician's discretion after all (see article 2). The prerogatives of the Hippocratic medical tradition are thereby still wisely preserved.

Distributive Justice

Next there is the question of the quality of justice in the delivery and distribution of health-care services. In other words, who shall get what portion of the health-care pie, given that there are limited medical resources to go around and many more people in need? As we saw in the immediately preceding section, the Hippocratic Oath pledges the doctor to remain free of all intentional injustice. Also, that same paragraph (P6) requires *equal treatment* of patients regardless of their sex or social status " . . . be they free or slaves." In addition, the physician pledges to abstain from any sexual relations with any of his patients (see P6, p. 69).

This personal commitment to equal treatment of the sick and fair play toward the weak and vulnerable expressed a supreme equalitarian ideal that was for the most part uncharacteristic of Greek and Roman thinking. It remains, I submit, a noble ideal by no means entirely realized in our own epoch. It expresses a humane value reaffirmed more broadly by the World Medical Association's Declaration of Geneva (see Appendix C, p. 193). This declaration of principles states, in part: "I will not permit considerations of religion, nationality, race, party politics, or social standing to intervene between my [medical] duty and any patient." Again, the robust foresight of the Hippocratic Oath and the progressive reinterpretation of some of its central values to better serve the contemporary medical setting, here on both a national and international scale, is unmistakable.

However, some important qualifications must be added. First, that the ancient Oath-taker admirably pledged to treat all of his patients equally (once he took them on) is not the same as saying that he charitably pledged to treat all those who sought his medical services. He could decline to treat whomever he wished. In fact, the now idealized notion that the high-minded physician has an additional duty to care for the sick, penniless, and down-trodden as part of an implicit debt owed to his community or to his God based on feelings of personal gratitude and mercy, was never a native Greek or Roman idea. Not until Judeo-Christian values took root during the late Roman Empire did this particular vision of charitable medical care for the needy and poor begin to take hold.[13] Even then, it was a practice especially encouraged by the advent of the early Christian hospitals which were relatively few in number.[14] Still, to say this is not to deny that the Pre-Christian doctors rendered their medical services free of charge from time to time. But such occurrences were infrequent in Greece and Rome and were usually calculated to enhance future business prospects or personal reputations.[15]

Second, as Robert Veatch has observed, the modern ideal of providing general access to medical services for all in need may directly conflict with fulfilling the familiar Hippocratic duty of doing what will most benefit the patient's health.[16] This constitutes a serious moral and economic dilemma. It is one which has not yet been fully acknowledged in the latest contemporary medical codes, nor was it even remotely anticipated by the Oath of Hippocrates.

The seriousness of the dilemma turns on the following consideration. As the limited supply, growing demand, and rising costs associated with many types of high-tech, life-sustaining therapies such as kidney dialysis, neo-natal intensive care, and open-heart surgery become increasingly parasitic on our nation's economy, this conflict of duty may well lead to a new form of economic triage according to which those who could not personally pay for certain costlier maladies would simply be economically doomed to die of them. In the United States, most citizens are painfully aware of the rising costs of medical care, and they also feel utterly helpless about reversing this trend. Some social commentators are beginning to ask whether in the future all who need access to various medical facilities and resources in order to remain alive and well, and who supposedly deserve the finest medical care whether rich or poor, will in fact be *equally* granted the treatment they so desperately require. Speaking for myself, I do not pretend to have the answer to this dilemma. But there is reason to think that a balanced solution can be reached only when the notion of prolonging or preserving human life at any cost is critically examined by every one of our major social institutions. Indeed, isn't this re-evaluation already well underway, for better or worse, in the present decade?

PROFESSIONAL COMMITMENT

In reference to my second claim, I said earlier that the Hippocratic Oath remains relevant to contemporary medicine by affirming the importance of the physician's covenant. As I see it, this covenant formally inaugurates the trusteeship relation between physician and patient on which all responsible medicine is ultimately founded. Yet it will be helpful to reconsider just what such an agreement means in modern terms.

Historically, it will be recalled that the second paragraph of the Oath served to: (a) bind the medical student to his teacher; (b) pledge the student to acquire and preserve the highest moral and technical standards of his art; and (c) require the student to remain vigilant so that the knowledge of his craft is extended only to those who are judged morally and intellectually fit (see P2, p. 69). Beyond this, a crucial social implication of the covenant was that it symbolized a public promise – what we would properly call a professional promise or commitment – to guard the welfare of patients along with the integrity of medicine itself against all major forms of abuse. This especially refers to forms of abuse that can originate from within the medical ranks. Such misconduct usually took the form of intellectual dishonesty or economic and moral wrongdoing. Today, the practice of medicine is also evidently vulnerable to these same human weaknesses. But most of these incidents have been vastly reduced since antiquity owing to the consistently higher caliber of medical students plus the rigorous standards imposed by both the state and by the profession on the entire medical enterprise.

Yet any such intellectual dishonesty, I imagine, included such deeds as these: regularly declining to seek consultations from medical peers though in serious doubt of a patient's proper diagnosis; losing touch with new therapeutic discoveries that might prove helpful while at the same time pretending to stay abreast of one's art; treating patients suffering from unfamiliar diseases, injuries, or complaints about which one knows little; and refusing to share with other practitioners effective cures of one's own that could help *their* patients regain health.

To be sure, economic and moral abuses are closely intertwined. Economic wrongdoing is almost always the direct or indirect result of some moral defect. Moreover, an assortment of intellectually dishonest behaviors of the sort described above are usually associated with even more fundamental moral deficiencies — notably lying (to oneself or others). Still, in order to further expand our analysis, it will be convenient to speak in an isolated way about a few of the more recurrent economic abuses that have plagued medicine over the centuries. These include: charging for services that are not in fact rendered; over-charging for services that are; concocting fictional diagnoses and remedies in order to secure extra business from unsuspecting patients who are really not ill; intentionally not curing patients as expeditiously as they might be cured in order to bilk them; and selling remedies or treatments that one knows in advance either will not work as described or will not work well enough to do any good. This lurid list of misanthropic acts is of course all quite illegal and universally regarded as unethical in our own day. In antiquity, however, this was not always so; medicine was largely unregulated and doctors were never licensed. Yet, even now, such abuses doubtless occur from time to time despite far stricter legal and professional codes.

It is a strange fact that medicine, by its very nature, is such an intimate enterprise that no finite account of its conceivable moral abuses could claim to be exhaustive or fully descriptive of its potential for wrongdoing. Nor need we consider in detail all the possible personal injuries or deaths that would be visited on innocent persons if individual doctors, or the societies to which they belonged, ever lost their moral footing. The medicine of Nazi Germany immediately comes to mind, but other historical examples could be cited as well.[17] On a purely individual scale, we saw that the Hippocratic Oath explicitly condemned as morally wrong having sexual relations with patients; treating patients unequally, based on their unequal social background or their sex; aiding in abortion or active euthanasia; gossiping about or otherwise spreading news of patients' personal affairs; and, in effect, intentionally committing any act which might inflict harm, mischief, or injustice of any sort.

The point is that this concise catalogue of moral prohibitions was originally intended to heighten the beginning doctor's awareness of, and commitment to, the noblest ideals at which medicine and its most trusted servants could aim. Currently, some may choose to quarrel about the particular details of the acts which the Hippocratic Oath condemns. And one can only applaud recent efforts toward a thoughtful reinterpretation and further refinement of the Oath's humanistic ideals. But few deny the admirable intent and relevance of the covenant itself, which finally transcends in importance any list of particular deeds. It stands as a public commitment to personal professional competence and moral rectitude. At the same time, it constitutes a solemn promise to one's professional peers to keep medicine free from scoundrels, charlatans, and crooks.

In the broadest sense, therefore, the covenant expresses the personal desire and resolve to make oneself and one's profession the very best that both can be. To be a professional, in this sense, is to profess competence and personal integrity not merely at the moment of entry into the ranks of trained healers, but every day of one's working life.

PRINCIPLES OF MEDICAL ETHICS
AMERICAN MEDICAL ASSOCIATION
REVISED, 1980

The medical profession has long subscribed to a body of ethical statements developed primarily for the benefit of the patient. As a member of this profession, a physician must recognize responsibility not only to patients, but also to society, to other health professionals, and to self. The following Principles adopted by the American Medical Association are not laws, but standards of conduct which define the essentials of honorable behavior for the physician.

   I. A physician shall be dedicated to providing competent medical service with compassion and respect for human dignity.

  II. A physician shall deal honestly with patients and colleagues, and strive to expose those physicians deficient in character or competence, or who engage in fraud or deception.

 III. A physician shall respect the law and also recognize a responsibility to seek changes in those requirements which are contrary to the best interests of the patient.

 IV. A physician shall respect the rights of patients, of colleagues, and of other health professionals, and shall safeguard patient confidences within the constraints of the law.

  V. A physician shall continue to study, apply, and advance scientific knowledge, make relevant information available to patients, colleagues, and the public, obtain consultation, and use the talents of other health professionals when indicated.

 VI. A physician shall, in the provision of appropriate patient care, except in emergencies, be free to choose whom to serve, with whom to associate, and the environment in which to provide medical services.

VII. A physician shall recognize a responsibility to participate in activities contributing to an improved community.

A PATIENT'S BILL OF RIGHTS
AMERICAN HOSPITAL ASSOCIATION
1973

The American Hospital Association presents a Patient's Bill of Rights with the expectation that observance of these rights will contribute to more effective patient care and greater satisfaction for the patient, his physician, and the hospital organization. Further, the Association presents these rights in the expectation that they will be supported by the hospital on behalf of its patients, as an integral part of the healing process. It is recognized that a personal relationship between the physician and the patient is essential for the provision of proper medical care. The traditional physician-patient relationship takes on a new dimension when care is rendered within an organizational structure. Legal precedent has established that the institution itself also has a responsibility to the patient. It is in recognition of these factors that these rights are affirmed.

1. The patient has the right to considerate and respectful care.
2. The patient has the right to obtain from his physician complete current information concerning his diagnosis, treatment, and prognosis in terms the patient can be reasonably expected to understand. When it is not medically advisable to give such information to the patient, the information should be made available to an appropriate person in his behalf. He has the right to know by name the physician responsible for coordinating his care.
3. The patient has the right to receive from his physician information necessary to give informed consent prior to the start of any procedure and/or treatment. Except in emergencies, such information for informed consent should include but not necessarily be limited to the specific procedure and/or treatment, the medically significant risks involved, and the probable duration of incapacitation. Where medically significant alternatives for care or treatment exist, or when the patient requests information concerning medical alternatives, the patient has the right to such information. The patient also has the right to know the name of the person responsible for the procedures and/or treatment.
4. The patient has the right to refuse treatment to the extent permitted by law, and to be informed of the medical consequences of his action.

5. The patient has the right to every consideration of his privacy concerning his own medical care program. Case discussion, consultation, examination, and treatment are confidential and should be conducted discreetly. Those not directly involved in his care must have the permission of the patient to be present.

6. The patient has the right to expect that all communications and records pertaining to his care should be treated as confidential.

7. The patient has the right to expect that within its capacity a hospital must make reasonable response to the request of a patient for services. The hospital must provide evaluation, service and/or referral as indicated by the urgency of the case. When medically permissible a patient may be transferred to another facility only after he has received complete information and explanation concerning the needs for and alternatives to such a transfer. The institution to which the patient is to be transferred must first have accepted the patient for transfer.

8. The patient has the right to obtain information as to any relationship of his hospital to other health care and educational institutions insofar as his care is concerned. The patient has the right to obtain information as to the existence of any professional relationships among individuals, by name, who are treating him.

9. The patient has the right to be advised if the hospital proposes to engage in or perform human experimentation affecting his care or treatment. The patient has the right to refuse to participate in such research projects.

10. The patient has the right to expect reasonable continuity of care. He has the right to know in advance what appointment times and physicians are available and where. The patient has the right to expect that the hospital will provide a mechanism whereby he is informed by his physician or a delegate of the physician of the patient's continuing health care requirements following discharge.

11. The patient has the right to examine and receive an explanation of his bill regardless of source of payment.

12. The patient has the right to know what hospital rules and regulations apply to his conduct as a patient.

No catalogue of rights can guarantee for the patient the kind of treatment he has a right to expect. A hospital has many functions to perform, including the prevention and treatment of disease, the education of both health professionals and patients, and the conduct of clinical research. All these activities must be conducted with an overriding concern for the patient, and, above all, the recognition of his dignity as a human being. Success in achieving this recognition assures success in the defense of the rights of the patient.

DECLARATION OF GENEVA
WORLD MEDICAL ASSOCIATION
AMENDED, 1983

At the time of being admitted as a member of the medical profession:

I solemnly pledge myself to consecrate my life to the service of humanity;
I will give to my teachers the respect and gratitude which is their due;
I will practice my profession with conscience and dignity;
The health of my patient will be my first consideration;
I will respect the secrets which are confided in me, *even after the patient has died*;
I will maintain by all the means in my power, the honor and the noble traditions of the medical profession;
My colleagues will be my brothers;
I will not permit considerations of religion, nationality, race, party politics or social standing to intervene between my duty and my patient;
I will maintain the utmost respect for human life from its beginning even under threat and I will not use my medical knowledge contrary to the laws of humanity.

I make these promises solemnly, freely, and upon my honor.

NOTES

CHAPTER I: THE STATUS OF THE PHYSICIAN

[1] Edelstein, Ludwig: 1967, 'The Distinctive Hellenism of Greek Medicine', *Ancient Medicine: Selected Papers of Ludwig Edelstein*, ed. by Lilian C. Temkin and Owsei Temkin, Johns Hopkins University Press, Baltimore, p. 376. See Homer *Odyssey* XVII. 383–85.

[2] *Ibid.*, Edelstein (trans.), p. 376.

[3] *Digesta* L. 13. 1; 3.

[4] See Edelstein, Ludwig and Emma J.: 1945, *Asclepius: A Collection and Interpretation of the Testimonies*, 2 vols., Johns Hopkins University Press, Baltimore. Also Ackerknecht, Edwin H.: 1955, *A Short History of Medicine*, Ronald Press, New York, p. 44.

[5] While the Edelsteins, above, n. 4, have done a remarkable and apparently exhaustive job of assembling (in volume one of their study) the known relevant evidence on the Asclepius religious cult, it must be acknowledged that in general the problems of tracing the continuity of medical developments from Homer's time to the fifth century B.C. are great. According to Sigerist, Henry E.: "We would like to know what rational medicine was like between Homer and Hippocrates – that is, in the four hundred years from the ninth to the fifth century B.C. – not that it has been lost, it was never written. The writing of medical books in the fifth century was a new phenomenon." See Sigerist, Henry E.: 1951 and 1961, *A History of Medicine*, 2 vols., Oxford University Press, New York, II, p. 85.

[6] Ackerknecht: *History of Medicine*, p. 44. See also Chap. VIII, n. 8 below.

[7] The actual sons of Asclepius were Machaon and Polalirius, both of whom Homer mentions for their bravery and service in the *Iliad*. In particular, it is Machaon who treats the wounds of Menelaus, adding further to the greatness of his family's healing skills. See *Iliad* IV. 192–218.

[8] *Iliad* XI. 514. Lattimore, Richard (trans.): 1961, *The Iliad of Homer*, University of Chicago Press, Chicago, p. 248.

[9] Aristotle *Politics* III.6.1282a. Rackham, H. (trans.): 1969, Loeb Classical Library, London, p. 227.

[10] Plato *Laws* IV. 720. Taylor, A. E. (trans.): 1966 *The Collected Dialogues of Plato*, ed. by Huntington Cairns and Edith Hamilton, Random House, New York, pp. 1310–11.

[11] *Ibid.*, *Laws* LX. 857c–e. For an alternative interpretation, see Fridolf Kudlien's monograph, 'Die Sklaven in der griechischen Medizin der klassischen und hellenistischen Zeit', *Forschungen zur antiken Sklaverei*, II, Steiner, Wiesbaden, 1968. Kudlien questions whether Plato's distinction is between free and slave physicians in respect to their *legal* status, or between physicians who are liberal or slavish-minded in respect to their *use* of the art.

[12] Temkin, Owsei: 1963, 'Greek Medicine as Science and Craft', *Isis* 33, p. 214.

13 Galen *De simplicium medicamentorum temperamentis ac facultatibus*. X. 22, cited by Edelstein: 'Distinctive Hellenism', p. 386.

14 Edelstein: 'The Hippocratic Physician', *Ancient Medicine*, p. 87.

15 Edelstein: 'Distinctive Hellenism', p. 385. For a comparison of the classical attitude toward work and the modern concept of the nobility of toil, see Michell, Humfrey: 1963, *The Economics of Ancient Greece*, 2nd ed., W. Heffer, Cambridge, p. 14. In general, there was a strong feeling on the part of the aristocrats of the fifth and fourth centuries (who constituted a vocal minority) that working with one's hands for a living was inferior to the cultivation of intellectual skills and pursuits done for their own sake. Edelstein comments ('The Ethics of the Greek Physician', *Ancient Medicine*, p. 326): "Completely free of any idealization of work as such, and considering it a dire necessity rather than an ennobling activity, the classical age judged all manual labor only by the standard of expertness and performance." Cf. Aristotle *Politics* 1337b 17, ff.

16 Amundsen, Darrel W.: 1978, 'History of Medical Ethics: Ancient Greece and Rome', *Encyclopedia of Bioethics*, ed. by Warren T. Reich, 4 vols., The Free Press, New York, III, p. 933. Hereafter I shall abbreviate references to the *Encyclopedia of Bioethics* as *EB*.

17 *Ibid.*, p. 933. On the topic of medical specialization in antiquity, Robert Flacelière recognizes a total of seven distinct health care practitioners of various sorts whom we would regard as medical professionals or para-professionals. These are (1) *the gymnastic master* (*paido-tribes*) "who was often a [2] *hygienist* and [3] *dietician* who could advise athletes on the best diet to follow . . . ," and could also reduce fractures, sprains, and dislocations; (4) *the pharmacist* (*pharmakopoles*) "who got his own supplies from the harvester of roots (*rhizotomos*);" (5) *oculists*, whose principal method of caring for their patients' eyes was using eye washes; (6) *dentists*, who knew how to fill teeth with gold; and (7) *the midwife*, one notable child of whom was Socrates. But it was unlikely that there exists within these divisions of labor a strict recognition of one's area of competence. Thus one must not be misled into reading contemporary medical specializations into these loosely defined ancient vocations. In addition to the seven medical functions just mentioned, there is some evidence that a few Greek city-states employed so-called *public physicians* (*iatros demosieuon*). However, Louis Cohn-Haft in his study 'The Public Physicians of Ancient Greece', *Smith College Studies in History*, XLII, Smith College, Northampton, Massachusetts, 1956, pp. 1–99, has raised serious doubts about the existence of such physicians in anything like a system of government-sponsored medical care. See Flacelière: 1967, 'Medicine', *Praeger Encyclopedia of Ancient Greek Civilization*, Praeger, New York, pp. 288–290.

18 Herodotus *The Persian Wars* II. 84. Rawlinson, George (trans.): 1942, *The Greek Historians. The Complete and Unabridged Works of Herodotus, Thucydides, Xenophon, and Arrian*, 2 vols., Random House, New York, II, p. 123.

19 According to John A. Wilson, these are the Edwin Smith Papyrus (surgical); the Embers Papyrus; the London Papyrus; the Hearst Papyrus; and the Berlin Papyrus. See Wilson: 1962, 'Medicine in Ancient Egypt', *Bulletin of the History of Medicine* 36, 116.

20 *Ibid.*, pp. 122–23.

21 Clifford Allbutt, T.: 1921, *Greek Medicine in Rome*, Macmillan and Co., London, p. 133. Allbutt's claim that the humoral doctrine originated in Egypt finds little support, and he supplies no evidence.

[22] Saunders, J. B.: 1963, *The Transitions from Ancient Egyptian to Greek Medicine*, University of Kansas Press, Lawrence, Kansas, pp. 15–33.

[23] For evidence of clear cases of Egyptian magic alongside more rational approaches to the treatment of disease, see Breasted, James H.: 1930, *The Edwin Smith Surgical Papyrus*, 2 vols., University of Chicago Press, Chicago, Case 9, p. 217, ff.

[24] Wilson: 'Ancient Egypt', p. 123. See also Edelstein: 'Distinctive Hellenism', pp. 376–77, and Temkin, Oswei: 1973, 'Health and Disease', *Dictionary of the History of Ideas*, ed. by Philip P. Wiener, 4 vols., Scribners and Son, New York, II, pp. 395–96.

[25] Amundsen, Darrel W.: 'History of Medical Ethics: Ancient Near East', *EB*, II, p. 882.

[26] *Ibid.*, p. 882.

[27] Aristotle *Politics* III. 1286a 12, ff. Jowett, Benjamin (trans.): 1941, *The Basic Works of Aristotle*, ed. by Richard McKeon, Random House, New York, p. 1198.

[28] Edelstein: 'Distinctive Hellenism', p. 375. Cf. Philemon's *Fragmenta comicorum greacorum* IV. 1841, cited by Edelstein, p. 375.

[29] There was no concept of "professions" in the classical age, as has been emphasized especially by Zimmern, Alfred E.: 1915, *The Greek Commonwealth*, 2nd ed., rev., Clarendon Press, Oxford, pp. 257, ff. It is probably true that in Hellenistic or later times, physicians as a class continued to acquire a greater social status; that their status varied according to different schools or individual authors like Plutarch or Galen, there can be no doubt. But on the whole, these were later developments. Kudlien rightly suggests that the ancients themselves held no unanimous opinion on the classification of medicine in the system of *technai*. He writes: "One well-known and important fact is that medicine was not included in the classical canon of liberal arts; however, this does not mean that it was never counted among them . . . " (p. 450). See his 'Medicine as a "Liberal Art" and the Question of the Physician's Income', *Journal of the History of Medicine* 31, No. 4, 1976, 448–59. See also n. 15, above.

[30] Edelstein: 'The Relation of Ancient Philosophy to Medicine', *Ancient Medicine*, p. 361.

[31] Frag. 31, Freeman, Kathleen (trans.): 1948, *Ancilla to the Pre-Socratics. A Complete Translation of the Fragments in Diels, Fragmente der Vorsokratiker*, Basil Blackwell, Oxford, p. 99. Aside from Democritus, it is quite possible that many of the Pythagoreans were themselves physicians. Evidence suggests that the Pythagorean recognition symbol, the pentagram, was also used as a symbol of health. Empedocles, too, was known as a healer. Hence early philosophical theories of the universe and early principles of medicine may have coalesced in these Pre-Socratic authors. See also Plato's *Gorgias* 464, ff. for allusive passages hinting at loose philosophical and medical links.

[32] Aristotle *Nicomachean Ethics* II. 1. 1104a 1–20. Ross, W. D. (trans.): *Basic Works*, pp. 9353–54.

[33] Jaeger, Werner: 1957, 'Aristotle's Use of Medicine as a Model in his Ethics', *Journal of the History of Science* 77, 54–61, especially 55–57. Cf. Gracia, Diego: 1978, 'The Structure of Medical Knowledge in Aristotle's Philosophy', *Sudhoffs Archiv* 62, No. 1, 17.

[34] Plato *Laches* 195c–d. Jowett (trans.): *Collected Dialogues*, p. 139.

[35] Plato *Republic* IV. 406, ff.

[36] Edelstein: 'Relation of Ancient Philosophy', p. 360. Cf. Aristotle *Eudaemonian Ethics* II. 2. 1227b 25, ff.

CHAPTER II: THEORIES OF HEALTH AND DISEASE

[1] Edelstein, Ludwig: 1970, 'Hippocrates', *The Oxford Classical Dictionary*, ed. by N. G. L. Hammond and H. H. Scullard, 2nd ed., Clarendon Press, Oxford, p. 518. Hereafter abbreviated *OCD*.

[2] Sigerist: *History of Medicine*, II, pp. 317–338, *passim*.

[3] Littré, Emile: 1839–1861, *Oeuvres completes d'Hippocrate*, 10 vols., Javal et Bourdeaux, Paris, I, p. 602, cited by Sigerist: *ibid.*, II, p. 318.

[4] Littré: *Oeuvres*, VII, p. 542, ff., cited by Sigerist: *History of Medicine*, II, p. 319.

[5] Sigerist: *History of Medicine*, II, p. 319.

[6] *Ibid.*, II, p. 320. Water was probably not as visible in the body (at least in its clear form) as were black discharges during severe illnesses. Hence the switch to black bile probably had an observational basis.

[7] *Ibid.*, II, p. 334. Sigerist points out that the Greek word *krasis* means not only "blending" but "temperature". In medical texts *krasis* usually translates "temperament".

[8] *Ibid.*, II, p. 321.

[9] *On the Nature of Man*, Littré, VI, p. 53, cited by Sigerist, *ibid.*, p. 14. In Gordon: 1945, *Medicine Throughout Antiquity*, F. A. Davis Co., Philadelphia, p. 514, Benjamin L. Gordon has pointed out that although as a rule *contraria contratis curantur* was recommended, it was not blindly followed by the Hippocratic physicians. See Hippocrates, *Ancient Medicine*, Chaps. I, XV.

[10] Hippocrates *On the Heart* VI. 6.

[11] Hippocrates *Epidemics* VI. 5, cited by Sigerist (trans.): *History of Medicine*, II, p. 326.

[12] Sigerist: *History of Medicine*, II, p. 328. Emphases added.

[13] Hippocrates *Epidemics* I. Littré, ed., II, p. 678, ff., cited by Sigerist, *ibid.*, II, p. 328.

[14] Hippocrates *Prognostics*, Littré, ed., II, p. 168, ff., cited by Sigerist, *ibid.*, II, p. 329. Emphasis added.

[15] Littre, ed.: *Oeuvres*, II, p. 598, cited by Sigerist, *Ibid.*, II, p. 331. Traditionally the medical school at Cnidus has been regarded by medical historians as having been more localistic in their description and classification of diseases than the school at Cos headed by Hippocrates. See Biggart, J. H.: 1971, 'Cnidus v. Cos', *Ulster Medical Journal* 41, 1–9.

[16] Frag. 234, Freeman (trans.): *Ancilla*, p. 112.

[17] Theophrastus *De sensu* 58 (DK 58, A 133).

[18] Robinson, John M. (trans.): 1968, *An Introduction to Early Greek Philosophy*, Houghton Mifflin, New York, p. 221.

[19] The atomic theory of Leucippus and Democritus, and the modifications thereto at the hands of Epicurus, formed the basis of a later medical sect known as the Methodists. See p. 30, and Moon, R. O.: 1909, *The Relation of Medicine to Philosophy*, Longmans, Green, and Co., New York, Chap. II.

[20] In the case of Plato, I am waiving the question of whether the medical theories found in his *Timaeus* were his own espoused views or those of a Pythagorean who is the speaker there.

[21] Plato, *Republic*, Shorey, Paul (trans.): *Collected Dialogues*, p. 687. Emphasis added. At *Timaeus* 81b–82b, Plato also distinguishes between decay (due to the natural aging

process) and disease. Decay is said to be the result of the wearing out of the elementary triangles without the intervention of disease. In contrast, disease is abnormal, morbid decay occasioned by the corruption (*phthora*) and reversal of the blood currents inside the body that normally foster growth. See Cornford, Francis M.: 1937, *Plato's Cosmology*, Bobbs-Merrill, New York, p. 336.

22 Plato *Republic* 445e, Shorey (trans.), *ibid*., p. 687.

23 At *Timaeus* 69d–72d, reason, spirit, and appetite are given location in the head, heart, and liver respectively.

24 *Timaeus* 87b, ff. It is interesting to note that *nosos*, which we commonly render "disease," is a Greek word that also means distress, affliction, *evil*. Liddell, H. G. and Scott, R.: 1972, *Greek English Lexicon* (Abridged), Clarendon Press, Oxford, p. 467.

25 *Timaeus* 29e–30b.

26 Gordon: *Medicine Throughout Antiquity*, p. 569, ff., states that Aristotle mentions over 500 different species of animals, and probably dissected at least 50. There is nothing in his writings to suggest that Aristotle dissected an adult human cadaver. He *may* have dissected human embryos, however.

27 See Solmsen, Friedrich: 1957, 'The Vital Heat, the Inborn Pneuma and the Aether', *The Journal of Hellenic Studies* 77, 119–23.

28 Aristotle *Metaphysics* 1014a 26, ff. and Kahn, Charles H.: 1960, *Anaximander and the Origins of Greek Cosmology*, Columbia University Press, New York, p. 120.

29 Aristotle *Physics* 246b 4, ff., Hardie, H. P. and Gay, R. K. (trans.): *Basic Works*, pp. 346–47. Emphases added.

30 Singer, Charles and Peck, Arthur: 'Zoology', *OCD*, p. 1140. On the part played by the four Aristotelian elements in determining the biological uniqueness of each species and members thereof, see Aristotle's *Parts of Animals* 645b, ff. Also see Hall, Thomas S.: 1974, 'Idiosyncrasy: Greek Medical Ideas of Uniqueness', *Sudhoffs Archiv* 58, No. 3, 283–302. Note that Aristotle regarded *the excellence of man's physical blend* as superior to all other animals as *proved* by man's higher range of intelligence, according to Singer.

31 Aristotle *Physics* I. 7–8.

32 Aristotle *Generation of Animals* V. 49. 784b 33, cited by Edelstein, trans., *Ancient Medicine*, p. 390.

33 Moon: *Relation of Medicine to Philosophy*, pp. 17–22, *passim*.

34 Gordon: *Medicine Throughout Antiquity*, p. 679.

35 *Ibid*., p. 703.

36 Neuburger, M. (trans.): 1910, *History of Medicine*, Oxford University Press, London, cited by Gordon, *ibid*., p. 706.

37 Penella, Robert J. and Hall, Thomas S. 1973, 'Galen's "On the Best Constitution of Our Body". Introduction, Translation, and Notes', *Bulletin of the History of Medicine* 47, No. 4, 283.

38 Gordon: *Medicine Throughout Antiquity*, p. 711. Like Plato, Galen attributes reason, emotion, and desire to the locations of the head (brain), heart, and belly respectively.

39 Galen *De sanitate tuenda* I. 4–5.

40 Galen *De facultatibus naturalibus* I. 6. This hierarchical structure of the body Galen largely adopted from Aristotle. Cf. Aristotle *On the Parts of Animals* 646b 12–27. Note that Aristotle in fact called like-parted things "homoeomerous."

41 Galen: 'On the Best Constitution', Penella and Hall (trans.), p. 294.

[42] Galen *De sanitate tuenda* II. 5.
[43] Galen: 'On the Best Constitution', Penella and Hall, (trans.), pp. 294–95. Emphases added.
[44] Galen *De temperamentis* I. 8.
[45] Galen: 'On the Best Constitution', Penella and Hall, (trans.), pp. 293–94. Emphases added.
[46] Kudlien, Fridolf: 1973, 'The Old Greek Concept of "Relative Health" ', *Journal of the History of Behavioral Sciences* 9, 58.
[47] Edelstein: 'The Dietetics of Antiquity', *Ancient Medicine*, pp. 313–14.
[48] Kudlien: 'The Old Greek Concept', p. 55.

CHAPTER III: ATTITUDES TOWARD DEATH

[1] Kalish, Richard: 'Attitudes Toward Death', *EB*, I, p. 286.
[2] Choron, Jacque: 1963, *Death and Western Thought*, Macmillan, New York, p. 32.
[3] Prophyrius *Vita Pythagorae* 19, cited by Robinson: *Early Greek Philosophy*, p. 57, and pp. 58–62, *passim*.
[4] Homer *Iliad* XVI. 856; XIV. 518. In Homer, *soma* is restricted to mean the dead body of man or beast; for a living body Homer uses the word *demas*. After Homer's time, *soma* referred to the body of either the living or the dead (as we also construe it). Liddell and Scott: *Greek Lexicon*, p. 688. The word *thymos* alone Homer associates with consciousness involving thought and feeling; *psyche* he reserves to designate life and movement. On this point see Peters, F. E.: 1967, *Greek Philosophical Terms*, New York University Press, New York, pp. 166–67. In general, see Snell, Bruno: 1953, *The Discovery of the Greek Mind*, trans. by T. G. Rosenmeyer, Harper Torchbook, Harper and Row, New York, Chap. I.
[5] Homer *Odyssey* XI. 488–491. Lattimore, Richard (trans.): 1967, *The Odyssey of Homer*, Harper and Row, New York, p. 180. Hesiod shows a brighter destiny for the heroes of old, cf. *Works and Days* 166.
[6] Cornford, Francis M.: 1967, *From Religion to Philosophy*, Harper, New York, p. xxiv, ff.
[7] Vermeule, Emily: 1979, *Aspects of Death in Early Greek Art and Poetry*, The Sather Classical Lectures, vol. XLVI, University of California Press, California, p. 41.
[8] *Ibid.*, p. 41.
[9] *Ibid.*, pp. 8–9. A fine example of the soul-bird is seen in an Attic white-ground lekythos vase of the fifth century. It depicts the boatman Charon, Hermes, and the soul-birds in Hades. This specimen was first brought to my attention by Philip Lockhart, and is photographically reproduced on p. 9 of Vermeule's book.
[10] *Ibid.*, p. 37.
[11] Choron: *Death and Western Thought*, p. 35.
[12] Frag. 2, Freeman (trans.): *Ancilla*, p. 19.
[13] Flew, Anthony: 1967, 'Immortality', *Encyclopedia of Philosophy*, ed. by Paul Edwards, 8 vols. Collier Press, New York, IV, p. 140. Emphasis added. Hereafter abbreviated *EP*.
[14] Choron: *Death and Western Thought*, p. 33.
[15] The loose sense of the word death is normally intended in such statements as "I

believe in life after death." If the word death did not bend in these contexts, a whole range of perfectly understandable utterances would have to be dismissed.

16 Heraclitus, Frags. 36, 48, 62, and 88, in Freeman's *Ancilla.*

17 Parmenides, Frags. B8 (lines 1–25; lines 36–40), in Diels, H. and Kranz, W.: 1903, *Die Fragmente der Vorsokratiker*, 11th ed., Weidmann, Berlin.

18 Cf. Empedocles Frags. 8 and 9 from 'On Nature' and Empedocles Frags. 146 and 147 from 'Purifications', in Freeman.

19 Choron: *Death and Western Thought*, p. 32.

20 *Ibid.*, p. 32.

21 Pindar, Frag. 137. Cf. Frag. 129, cited and trans. by Mair, A. W.: 'Life and Death: Greek', *Encyclopedia of Religion and Ethics*, ed. by James Hastings, 13 vols., Charles Scribner & Sons, New York, 1852–1922, VIII, p. 30. Hereafter abbreviated *ERE*.

22 Sophocles *Antigone* V. 332; 360. Roche, Paul (trans.): 1958, *The Oedipus Plays of Sophocles*, Mentor, New York, cited by Choron, *Death and Western Thought*, p. 42.

23 Bowra, C. M.: 1959, *The Greek Experience*, Mentor, New York, p. 49.

24 Roche, Paul (trans.): cited by Choron, *Death and Western Thought*, pp. 42–43.

25 Aeschylus *Agamemnon* 1148, ff. Mair, A. W. (trans.): 'Life and Death (Greek)', *ERE*, VIII, p. 30.

26 See also Chap. VII.

27 For more on these four points, see Dover, K. J.: 1974, *Greek Popular Morality in the Time of Plato and Aristotle*, University of California Press, Los Angeles, pp. 243–46. See also Kurtz, Donna C. and Boardman, John: 1971, *Greek Burial Customs*, Cornell University Press, Ithaca, New York, p. 206, and on ghosts in Classical times, Rohde, E.: 1925, *Psyche*, Hillis, W. B. (trans.), 8th ed., Paul Treuch, Trubner Ltd., London, pp. 593–95.

28 Dover: *Greek Popular Morality*, p. 245. Cf. Plato's *Laws* X.

29 Homer *Odyssey* XI. 72, ff.

30 Rose, Herbert J.: 'Disposal of the Dead', *OCD*, p. 314. On pollution and purification in connection with homicide, cf. Macdowell, D. M.: 1963, *Athenian Homicide Law in the Age of the Orators*, University of Manchester Press, Manchester, Chaps. I, XII, and XIV.

31 Cf. *Tim* 29d and *Phaedo* 107b–c. See also Plato's *Letter* VII 335a–c.

32 Tredennick, Hugh (trans.): *Collected Dialogues*, p. 52.

33 For Plato's standard definition of death, *Phaedo* 64c–d; for a physical account of decaying and death, *Timaeus* 80c–d and n. 21, Chapter II, above; for his views on reincarnation, *Republic* X.

34 I am inclined to agree with those scholars who attribute to the historical Socrates the agnostic views on immortality at *Apology* 38c, ff., and to Plato alone the doctrine of immortality supported in *Phaedo* and elsewhere. See Chap. VII.

35 Aristotle: *De Anima* 412b 5; 413a 1.

36 *Ibid.*, *De Anima* 407b 22.

37 *Ibid.*, *De Anima* 415a 27, ff., Smith, J. A. (trans.): *Basic Works*, p. 561. Cf. Plato's *Symposium* 206–208 for a similar perspective expressed by Diotima. Emphases added. For a discussion of dissenting views, see Copleston, Frederick: 1962, *A History of Philosophy*, Image, New York, I, P. 2, pp. 70–73.

38 Aristotle *Rhetoric* 1382a 20–27. Roberts, W. Rhys (trans.): *Basic Works*, p. 1389.

39 Aristotle *Nichomachean Ethics* III. 6. 1115a 26–35. Emphases added.

[40] Aristotle *De Caelo* I. 4. 271a 33.

[41] This last claim, called hylozoism, was similarly endorsed by a number of the Pre-Socratics.

[42] DeLacy, P. H.: 'Epicurus', *EP*, II, p. 5.

[43] Diogenes Laertius *De vitis* X. 138. The caveat against untherapeutic philosophers derives from *The Stoic and Epicurean Philosophers*, ed. by Whitney J. Oates, Random House, New York, 1940, p. 31, cited by Choron, *Death and Western Thought*, p. 58.

[44] DeLacy, 'Epicurus', *EP*, II, 4–5.

[45] 'Letter to Menoecus', Oates, *Stoic and Epicurean*, pp. 30–1.

[46] For a possible rejoinder, see Miller, Fred: 1976, 'Epicurus on the Art of Dying', *The Southern Journal of Philosophy* 14, No. 2, pp. 169–77, *passim*.

[47] For Stoic views on death, see p. 121, below.

CHAPTER IV: WHO WAS HIPPOCRATES?

[1] Amundsen: 'History of Medical Ethics: Ancient Near East', *EB*, II, p. 883.

[2] For a look at the contribution of Near Eastern art to later developments in Egyptian and Greek art, see Gardner, Helen: 1975, *Art Through the Ages*, ed. by Horst de la Croix and Richard G. Taney, Harcourt Brace, New York, pp. 104, 126. On Egyptian antecedents and possible influences on Greek medicine, see Saunders, J. B.: *Transitions from Egyptian to Greek Medicine*, p. 1, ff.

[3] Amundsen: 'Ancient Near East', *EB*, II, p. 883.

[4] Aristotle *Politics* III. 15. 1286a 12, ff., and Chap. I, p. 9, above.

[5] All the preceding features of the Code of Hammurabi are discussed by Amundsen: 'Ancient Near East', *EB*, II, p. 882. See also Biggs, Robert: 1969, 'Medicine in Ancient Mesopotamia', *History of Science* 8, 94–105.

[6] Amundsen: 'Ancient Near East', *EB*, II, p. 883.

[7] *Dinkard* 157. 19. Ratanshan Erachshah Kohiyar (trans.): 1874–1891, ed. by Peshotan dastur Behramjee Sanjana, *The Dinkard: The Original Pahlavi Text*, Duftur Ashkara Press, Bombay, cited by Amundsen, *ibid.*, p. 883.

[8] See the Hippocratic treatise *The Sacred Disease* in Jones, W. H. S. (trans.): 1962, *Hippocrates*, 4 vols., Harvard University Press, Cambridge, Massachusetts, II, pp. 139–41. Prior to the emergence and influence of the Pre-Socratic philosophers, there was a time when diseases were presumed by the Greeks to be sent from the gods, as they were by the Near Eastern cultures. Homer restricted divine causes mainly to epidemic diseases. See *Iliad* I. 8–10. However, Hesiod appears to have been less restrictive and thought that the gods brought all diseases upon man. Cf. *Works and Days* 103–105.

[9] Amundsen: 'Ancient Near East', *EB*, II, p. 881.

[10] *The Edwin Smith Papyrus* 6, 8a, 20; and *The Papyrus Ebers* 108–110.

[11] Amundsen: 'Ancient Near East', *EB*, II, p. 881.

[12] *Ibid.*, II, p. 882. See *The Vendidad* 15. 2. 12–14. Darmesteter, James (trans.): 1880–1883, *The Zend-Avesta*, 3 vols., Clarendon Press, Oxford, and reprinted by Greenwood Press, Westport, Conn., 1972.

[13] Amundsen: 'Ancient Near East', *EB*, II, p. 882. This law Amundsen has located in the Middle Assyrian Laws, A 53.

[14] Plato *Protagoras* 331b–c.

[15] Plato *Phaedrus* 270b–d.

[16] Plato *Phaedrus* 270b, ff. Hackforth, R. (trans), *Collected Dialogues*, p. 515. The sense of the word "empirical" Socrates here intends is, of course, the superficial sense of ignoring the true causes or natures of things, which, on his view, lie beneath the surface of mere appearance.

[17] Aristotle *Politics* 1326a 15.

[18] Lyons, Albert S. and Joseph Petrucelli, R: 1978, *Medicine: An Illustrated History*, Narry N. Abrams, Inc., New York, pp. 207—10. The Roman coins in question are now housed in the British Museum, London.

[19] It is tempting to speculate on whether Hippocrates' theory of the etiology of disease was influenced by some features of the Egyptian theory of putrefaction. See above, Chap. I. This possibility cannot be ruled out. But just how and where Hippocrates may have gained knowledge of Egyptian medicine is a mystery. He did travel within Greece, we know, but that he visited Egypt we have no firm reason to believe. For ancient legends to the contrary and their likely misleading implications, see Sigerist: *History of Medicine*, II, p. 271.

[20] Edelstein: 'Hippocrates', *OCD*, p. 518. The best known of the ancient biographies of Hippocrates is *Life of Hippocrates*, by Soranus of Ephesus. The authorship of this work, however, is in doubt. For a useful evaluation of the ancient and medieval biographies on Hippocrates, see Sigerist: *History of Medicine*, II, p. 268, ff.

[21] Here I concur with Edelstein, whose explanations, with a few qualifications, has won the acceptance of Sigerist and several other leading classicists and medical historians. See Edelstein: 1931, *Peri aerov und die Sammlung der hippokratischen Schriften*, Weidmann, Berlin, p. 116, ff., cited by Sigerist, *ibid.*, II, pp. 266—67.

[22] Sigerist: *History of Medicine*, II, p. 267.

[23] *Ibid.*, p. 267. See Diller, H.: 1933, 'Zur Hippokratesauffassung des Galen', *Hermes* 68, 167—81.

[24] Just how one goes about counting up the manuscripts belonging to the Hippocratic Collection is a matter of current controversy. According to Sigerist, almost all the books belonging to the collection, with a few exceptions, date from the fifth and fourth centuries B.C. As I have argued, the bulk of the *Corpus Hippocraticum* represents all or most of the best medical literature from this earlier period that the Alexandrians could find. Sigerist comments: "There has been much speculation on the number of treatises included in the Collection, but this is a vain speculation for, unlike the Hermetic books of Egypt, the *Corpus Hippocraticum* was never firmly constituted, never divided into a set number of books, never rigidly fixed . . . . Still, the best Greek manuscripts contain about sixty books which were transmitted from the fifth and fourth centuries." Sigerist: *History of Medicine*, II, pp. 274—75. While Sigerist takes the position that at least *some* of the works in the collection may be genuine works by the historical Hippocrates, Edelstein holds that none of the books preserved under the name Hippocrates is genuine. Edelstein: 'Hippocrates', *OCD*, p. 518. For more on the problem, see Edelstein: 'The Genuine Works of Hippocrates', *Ancient Medicine*, pp. 133—44.

CHAPTER V: THE HIPPOCRATIC OATH

[1] Edelstein (trans.): 'The Hippocratic Oath', in *Ancient Medicine*, p. 6. For the Greek text, see *Hippocratis Opera*, ed. by I. L. Heiberg, Corpus Medicorum Graecorum, I, 1, pp. 4—5. The text is also reproduced in full by Edelstein, *ibid.*, p. 5.

² For the general evidence on the prevalence of abortion and suicide in the ancient world, see Crawley, A. E.: 'Foeticide', *ERE*, VI, p. 54b, ff.; and Mair, A. W.: 'Suicide', *ERE*, XII, pp. 31–2; also, Edelstein: 'Hippocratic Oath', pp. 9–20. A full discussion of the prevalence of abortion, euthanasia, and suicide in Greek times is addressed in Part Three.

³ Scribonius Largus *Professio medici.*

⁴ Nittis, Savas: 1940, 'The Authorship and Probable Date of the Hippocratic Oath', *Bulletin of the History of Medicine* 8, 1020.

⁵ See Thucydides *The Peloponnesian War* VII, 67 and 97; and Nittis.

⁶ Nittis: 'Authorship of the Hipoocratic Oath', p. 1020. On the question of the dramatic date of the *Protagoras*, see Vlastos, Gregory: 1957, introduction to Plato's *Protagoras*, ed. by G. Vlastos, Bobbs-Merrill, New York. On the *Phaedrus*, see More's article in *Patterns in Plato's Thought*, ed. by Jules M. Moravcsik, Synthese Historical Library, No. 6, Kluwer, Boston, Hingham, Massachusetts, 1973.

⁷ Nittis: 1939, 'The Hippocratic Oath in Reference to Lithotomy: A New Interpretation with Historical Notes on Castration', *Bulletin of the History of Medicine* 7, 721.

⁸ Littré: *Oeuvres*, IV, pp. 616–17.

⁹ Jones, W. H. S.: 1924, *The Doctor's Oath*, Cambridge Press, London, p. 48; and Edelstein: 'Hippocratic Oath', p. 27.

¹⁰ Nittis: 'Oath in Reference to Lithotomy', p. 721.

¹¹ According to Liddell and Scott: *Greek Lexicon*, p. 213, *ekchorreso* ordinarily means (1) to go out and away; (2) to slip out of or be dislocated; and (3) to give way, to give place to.

¹² Edelstein: 'Hippocratic Oath', p. 27, n. 80.

¹³ Sigerist: *History of Medicine*, II, p. 303. Also Amundsen, Darrel: 1978, 'The Physician's Obligation to Prolong Life: A Medical Duty without Classical Roots', *Hastings Center Report*, p. 26.

¹⁴ Aristoxenus 58d 1, Edelstein, trans., 'Hippocratic Oath', p. 30.

¹⁵ Plato *Timaeus* 87c, ff., especially 89d.

¹⁶ Edelstein: 'Hippocratic Oath', p. 31.

¹⁷ *Ibid.*, p. 32. Some commentators suggest that distress from kidney or bladder stone was a contributing cause of suicide, so painful are its unrelieved effects.

¹⁸ *Ibid.*, p. 32. Emphasis added.

¹⁹ Jones (trans.): *Hippocrates*, IV, p. 217.

²⁰ This alternative is explored in Kudlien, Fridolf: 1970, 'Medical Ethics and Popular Ethics in Greece and Rome', *Clio Medica* 6, 103–04.

²¹ Frag. 58, Freeman (trans.): *Ancilla*, p. 28.

²² Edelstein: 'Hippocratic Oath', p. 44. Edelstein's evidence here is that the Pythagorean Epaminondas is said to have honored his teacher as a father. See Diodorus X. 11. 2.

²³ DeVogel, C. J.: 1966, *Pythagoras and Early Pythagoreanism*, Royal VanGorcum Ltd., The Netherlands, p. 240. See Plato's *Symposium* for related cases of personal devotion, if not adoption.

²⁴ Edelstein: 'Hippocratic Oath', p. 26. Emphases added.

²⁵ Iamblichus *De Vita Pythaegorica* 163. DeVogel (trans.): *Pythagoras*, p. 234.

²⁶ The prominent role played by carefully calculated dietary therapies was suggested above, Chap. II, in connection with the *principle of contraries*. The Greeks tried to identify which foods engendered the qualities hot, cold, wet, and dry in order to restore

humoral balance to the sick patient. The Hippocratic author of *Regimen* I. 2, states that knowledge of the qualities of all foods and drink must be acquired by anyone aspiring to practice good medicine.

[27] *Epidemics I* XI. Jones (trans.: *Hippocrates*, I, p. 165.

[28] Edelstein: 'Hippocratic Oath', p. 22. Cf. Plato *Republic* I, concerning the crafts — including medicine — and the proper objects at which they aim.

[29] *Ibid.*, p. 35. In saying that the Pythagoreans alone judged sexual relations in terms of justice, Edelstein means that such conduct was *not* merely judged in accord with what was allowed by human laws or conventions. Justice here refers to a higher moral principle, ultimately mathematical in nature, that transcends human law.

[30] DeVogel: *Pythagoras*, pp. 237–38.

[31] To be unfaithful to one's spouse was to live in disharmony with nature (reason). The Stoics counted neither pleasure a good, nor pain an evil. Thus one was expected to endure the possible hardships of marriage.

[32] Plato *Laws* VIII. 842a–848c; also *Laws* I. 636c. Cf. Socrates' attitude according to Alcibiades' speech in the *Symposium*, and what DeVogel has rightly inferred is Plato's implicit disapproval of homosexuality there.

[33] Jones (trans.): *Hippocrates*, II, p. 313. Emphases added.

[34] Edelstein's evidence for these claims include Diogenes Laertius VIII. 15 (Aristoxenus) and a fragment from Aexis in Diels and Kranz: *Die Fragmente* (5th ed.), p. 464.

[35] In contrast with the Pythagorean's motivation for cultivating silence, which ultimately had a moral basis going beyond good manners, we read in the Hippocratic work *The Physician* (also discussed, below, p. 91): "The prudent man must also be careful of certain moral considerations — not only to be silent, but also of a great regularity of life, *since thereby his reputation will be greatly enhanced*." Jones (trans.): *Hippocrates*, II, pp. 311–12.

[36] Kudlien: 'Medical Ethics and Popular Ethics', p. 110. Kudlien cites as evidence for these two points Ziebarth, in *Pauly-Wissowa Realencyclopadie*, (1905 ed.) II, p. 2080; and Diogenes Laertius I.3.69 and I.7.98.

[37] Edelstein: 'Hippocratic Oath', pp. 15–18, *passim*.

[38] Plato *Republic* I. 332d–4. Shorey, Paul (trans.): *Collected Dialogues*, pp. 581–82. Emphases added. That physicians were ready targets for the charge of harming their patients is frequently commented on in the Hippocratic Collection. See, e.g., *The Art* 7; *Decorum* 14, 17; *On Joints* 67.

[39] Plato *Laws* XI. 932e, ff., Saunders, Trevor J. (trans.): 1970, *The Laws*, Penguin Books, New York, p. 480.

[40] *Ibid.*, p. 480. Emphases added.

[41] Edelstein: 'Hippocratic Oath', p. 10, n. 11.

[42] See above, p. 78.

[43] This possibility is also discussed by Kudlien, who, along with Littré before him, basically supports my wider reading of the proviso against supplying poison. Cf. Kudlien: 'Medical Ethics and Popular Ethics', p. 97, ff., and Littré: *Oeuvres*, IV, pp. 662, ff.

[44] Amundsen, Darrel W.: 1974, 'Romanticizing the Ancient Medical Profession: The Characterization of the Physician in the Greco-Roman Novel', *Bulletin of the History of Medicine* 48, No. 3, 323, n. 14.

[45] This claim is admittedly open to debate. Edelstein contends ('Hippocratic Oath', p. 10) that the Greeks were confident that they could detect such crimes as poisoning

by getting a confession through torturing prime suspects, if need be. Louis Lewin discusses ancient methods of investigating alleged cases of poisoning in his *Die Gifte in der Weltgeschichte*, J. Springer, Berlin, 1920, pp. 37, ff. Kudlien concurs with my judgment, against Edelstein, that the Greeks during this period possessed neither a pathological anatomy nor a legal medicine in the modern sense ('Medical Ethics and Popular Ethics', p. 102).

[46] This very topic of the doctor's moral responsibility (or lack thereof) is dramatized by the second century A.D. Roman writer Apuleius, in his novel *Metamorphoses* (also known as *The Golden Ass*). Briefly, the story is told of a physician who is approached with a large sum of money for the sale of a lethal poison by a third party. This third party claims the poison is for a sick friend who is described as suffering terribly from an incurable disease, and wishes to quit life. Luckily, the physician guesses that this poison is intended for a murder plot of some sort. He substitutes for it a sleeping compound, thereby sparing the victim's life. Here, I think, Apuleius strongly hints at the vulnerability of the physician to witting or unwitting involvement in murder.

[47] Some noteworthy alternative translations include W. H. S. Jones' (*Hippocrates*, I, p. 299) "Similarly I will not give to a woman a pessary to cause abortion"; John T. Noonan's ('An Almost Absolute Value in History', in *The Morality of Abortion*, Harvard University Press, Cambridge, Mass., 1970, p. 4.) "Similarly I will not give to a woman an abortifacient pessary"; and Benjamin L. Gordon's (*Medicine Throughout Antiquity*, p. 517) "[A]nd in like manner I will not give to a woman a pessary to produce abortion." But John Chadwick and W. N. Mann (*The Medical Works of Hippocrates*, Blackwell Scientific Publications, Oxford, 1950, p. 9) join Edelstein as non-literalists just here, translating: "Neither will I give a woman means to produce an abortion." So, too, are Charles Singer and E. Ashworth Underwood (*A Short History of Medicine*, 2nd ed., University Press, Oxford, 1962, p. 32), who prefer: "[A]nd especially I will not aid a woman to procure abortion."

[48] For example, the first century A.D. writer Scribonius Largus took the Hippocratic Oath to forbid abortion *simplicitur* (see Chap. VIII, below). So did his near contemporary Soranus, who quotes the Oath as saying "I will give to no one an abortive." Soranus, incidentally, disagreed with the Oath's categorical restriction (Chap. VI, below).

[49] Liddell and Scott, *Greek–English Lexicon* (Cambridge, 1972), p. 487.

[50] See Chap. VI, pp. 107–108, below.

[51] The Greek word *pessos* meant a "medicated plug of wool or lint to be introduced into the vagina, anus, etc., pessary," according to Liddell and Scott's *Greek Lexicon*, 9th ed., p. 1396. Among the few ancient authors who mention the pessary are the botanist Theophrastus (fl. 340 B.C.) in his *History of Plants* 9.20.4. He cites a plant called birthwort (*Aristolochia rotunda*) as good "for the womb as a pessary: . . . in cases of prolapsed uterus it is used in water as a lotion." *Inquiry into Plants*, Arthur Hort, (trans.), Loeb Classical Library, Harvard University Press, Cambridge, Mass., II, p. 319. Another source is the medical writer Celsus (fl. 25 A.D.). In his *de Medicina* V. 21–22, he describes in detail various kinds of pessaries useful (a) for inducing menstruation; (b) "to mollify the womb"; (c) "against inflammations of the womb"; (d) to render the expulsion of an already dead fetus; (e) to control hysterical fits "owing to genital disease"; and (f) to promote conception. Oddly, no mention is made of the abortifacient pessary by these authors. But see Soranus' *Gynecology* I.65, where several such recipes are provided. Cf. Dioscorides Medicus I.106, II.61.

52 Soranus does not mention death to the mother caused by a pessary. He does, however, warn that "one should choose those which are not too pungent, that they may not cause too great a sympathetic reaction and heat." *Soranus' Gynecology*, Owsei Temkin (trans.), Johns Hopkins Press, Baltimore, 1956, p. 67.

53 The respect for human life principle is discussed in detail below, Chap. VI.

54 Frankena, William K.: 1975, 'The Ethics of Respect for Life', *Respect for Life: In Medicine, Philosophy, and the Law*, ed. by Stephen Barker, Johns Hopkins University Press, Baltimore, p. 38.

55 Kudlien: 'Medical Ethics and Popular Ethics', pp. 109–10. Kudlien introduces the testimony of M. Nilsson, according to whom one need not search for Pythagorean influences to explain satisfactorily those ethical rules which were common to a religiously tinged stock of popular Greek beliefs. See Nilsson: 1961, *Geschichte der griechischen Religion*, 2nd ed., n.p., München, II, p. 130.

56 DeVogel: *Pythagoras*, p. 241. Another objection that might be raised against Edelstein's contention that the Oath is of exclusively Pythagorean origin, centers on a linguistic problem. The problem is that the Pythagoreans originally thrived in Dorian-speaking Greek communities on the more western Mediterranean shores of Italy. But since the Oath is written in Ionic Greek – not in the Pythagorean-favored Doric dialect – this fact seems to further complicate Edelstein's thesis.

57 Liddell and Scott: *Greek Lexicon*, p. 289; p. 303.

58 Jones: *Hippocrates*, II, p. 259.

59 Hippocrates *Law* I. Jones (trans.): *Hippocrates*, II, p. 263.

60 *Ibid.*, II, p. 265.

61 Compare the Hippocratic Oath's opposition to the practice of surgery discussed above, pp. 72–76. *The Physician* thus contains another example of logical inconsistencies between The Oath and the Hippocratic Collection which point to multiple authorship. Additional inconsistencies regarding abortion and euthanasia will be treated in Part Three.

62 On the topic of the development of the science of human anatomy in Greco-Roman times, see Edelstein: 'The History of Anatomy in Antiquity', *Ancient Medicine*, pp. 247–301.

63 *The Physician* I. Jones (trans.): *Hippocrates*, II, pp. 311–13.

64 For one such attempt, see Edelstein: 'The Professional Ethics of the Greek Physician', in *Ancient Medicine*, p. 329, ff.

65 *Precepts* VI. Jones (trans.): *Hippocrates*, I, p. 319. Emphases added. For a related passage on fee-setting which evidences a clear regard for the patient's welfare in relation to the hardship imposed by illness, see *Precepts* IV.

66 *Decorum* V. Jones (trans.), *ibid.*, II, p. 287.

67 *Decorum* XIII. Jones (trans.), *ibid.*, II, p. 295.

68 *Decorum* XIV. Jones (trans.), *ibid.*, II, p. 297.

69 *Decorum* XII. Jones (trans.), *ibid.*, II, p. 295.

70 See the Oath, above, P6, p. 69. Emphasis added.

71 This is not to say that the actual taking of craft-related oaths was rare, however, since they may have been kept secret through an oral tradition of instruction. Among Greek guilds, Marcus N. Tod singles out three of the more common types: (1) *religious guilds*, particularly those devoted to the Dionysiac art forms of music and drama; (2) *wholesale merchant associations*, devoted to religious, social, and commercial ends; and (3) *crafts*

guilds, composed of workers of the same craft, industry, or trade such as launderers, tanners, doctors, cobblers. Of the last type, Tod states: "Their main function was religious and social rather than economic . . . " Tod further states that, although these three types of guilds were numerous and very popular during both Greek and Roman times, there are few literary references to them throughout antiquity. See Tod's article 'Greek Clubs', *OCD*, pp. 254–555, which contains a useful bibliography.

[72] *Precepts* VI. Jones (trans.), *ibid.*, I, p. 319.

CHAPTER VI: THE PROBLEM OF ABORTION

[1] It appears certain that Galen wrote no commentary on the Hippocratic Oath, and he nowhere mentions the Oath by name. What his personal views were on abortion, infanticide, or euthanasia is highly conjectural. But since he much admired Aristotle's ethical theories, it is possible that he followed Aristotle pretty closely on these matters, too. Norman K. Himes, in his *Medical History of Contraception*, Williams and Wilkins Co., Baltimore, 1936, p. 75, reports that one " . . . Dr. Moissides of Athens informs me that he has searched the works of Galen and has been able to find no discussion of contraception." Similarly, I am unaware of any direct evidence of Galen's views on abortion.

[2] Typical of such a commitment to the relation between doing good and fairing well, is Plato at *Republic* X.621d. There is an instructive discussion of eudaimonism in Vernon J. Bourke: 1970, *History of Ethics*, 2 vols., Image Books, New York, I, Chaps. I–III. See also MacIntyre, Alasdair: 1976, *A Short History of Ethics*, Macmillan, New York, Chaps. VII–VIII, for a useful comparison of ancient eudaimonism and contemporary moral theory.

[3] Lecky, William E.: 1905, *History of European Morals: From Augustus to Charlemagne*, 2 vols., Longmans, Green, and Co., London. See especially II, p. 19, ff.

[4] Westermarck, Edward A.: 1906–08, *The Origin and Development of the Moral Ideas*, 2 vols., Macmillan, London. See especially I, Chap. XVII, pp. 393–413, entitled 'The Killing of Parents, Sick Persons, Children – Feticide'.

[5] Durant, Will: 1939, *The Story of Civilization*, 10 vols., Simon and Schuster, New York, II and III.

[6] The exception here was the Greek city of Thebes. It is conjectured that the Pythagorean, Philolaus, who was known to have advocated a strict respect for human life ethic, lived in Thebes and succeeded in influencing that city's legislation which barred, according to legend, both infanticide and suicide. The principal evidence for such laws is to be found in Aelian's *Varia Historiae* II. 7, cited by Lecky: *European Morals*, II, p. 26. Yet there is reason to believe that the Theban law referred only to healthy, not deformed, infants. The crime was punishable by death in any case.

[7] In 318 A.D., the Christian Emperor Constantine, falling under the influence of the Judeo-Christian respect for human life ethic, declared the act of fathers slaying their sons or daughters at any age to be a crime. According to Langer, William L.: 1974, 'Infanticide: A Historical Survey', *History of Childhood Quarterly: The Journal of Psychohistory* 1, 355, Constantine never explicitly forbade infanticide by name though presumably he meant to include it. But infanticide was made a crime punishable by death toward the end of the fourth century by the Emperors Valentinian, Valens, and

Gratian. All this must be understood in the context of ancient Roman legal powers, extending back to the early Roman Republic, which gave the *paterfamilias* the legal power (*patria potestas*) of life and death over members of his household, especially his children and slaves. Similar laws or customs were ascribed to Greek fathers during the Archaic period. Langer, commenting on infanticide and the laws of the late Roman Empire (p. 363, n. 8), notes that even the Christian emperors were somewhat ambivalent. For example, they did nothing to abolish exposure (a passive form of infanticide) or the sale of exposed children (by those who rescued such children) as slaves. In fact it was not until the time of Justinian in 529 A.D., that the subjection of foundlings to slavery was outlawed.

8 This is not to imply that there were *no* voices to be heard during pagan times in opposition to abortion or infanticide. Among those writers from the Roman period who described the widespread practice of abortion with disapproval were Ovid (*De nuce* 22–23); Seneca (*Ad Helvia* XVI); Favorinus (quoted by Aulus Gellius in his *Noctes Atticae* 12.1); and Juvenal (*Sat.* VI. 592). At Athens during the Golden Age, both custom and law provided that on or before the *tenth day* of its new life, the infant was formally accepted into his family in a religious ritual around the hearth, receiving presents from well-wishers and a name. Parents who thereafter exposed such children to die were punished by the state. See Durant: *Civilization*, II, pp. 287–88 and Westermarck, *Moral Ideas*, I, p. 410. In Hellenistic times, the Jewish philosopher Philo Judaeus (c. 30 B.C.–A.D. 45) denounced the exposure of children on religious grounds, calling it murder. Early Christian writers like Tertullian (c. A.D. 160–c. 240), an active proponent of the Judeo-Christian respect for human life religious ethic, condemned not only infanticide but also abortion as morally wrong. For he construed the developing fetus as a human life, declaring: "He also is a man who is about to be one. Even every fruit already exists in its seed" (*Apologeticus* 9).

9 This was particularly a motive of some upper class Hellenistic and Roman women. "Seneca thinks Helvia worthy of special praise because she has never destroyed her expected child within her womb 'after the fashion of many other women, whose attractions are to be found in their beauty alone,' " states Westermarck: *Moral Ideas*, I, p. 416. He quotes Seneca from *Ad Helvia* XVI. See also Durant: *Civilization*, II, p. 567.

10 Writing on infanticide in antiquity, Langer suggests that "modern humanitarian sentiment makes it difficult to recapture the relatively detached attitude of the parents towards their offspring. Babies were looked upon as the unavoidable result of normal sex relations, often as an undesirable burden rather than as a blessing." 'Infanticide', p. 355. While this may be *part* of the story, I think Langer may mislead his readers into thinking that little warmth or compassion was commonly felt and expressed by Greek and Roman parents toward their young. This is, of course, an unwarranted inference. Once parents decide to keep their newborn (a decision customarily made within a day or two of its birth), they are standardly raised with a great deal of indulgence and affection.

11 This was no doubt a recurring motive for both abortion and infanticide. Lecky (*European Morals*, II, p. 28, n. 1) states that "Quintilian [fl. 60 A.D.] speaks in a tone of apology, if not justification, of the exposition of the children *of destitute parents* (*Declamationes* 306) and even Plutarch speaks of it without censure (*De Amor. Prolis.*)." Emphases added.

12 The Greeks felt the pressures of population growth more than the Romans. The Greek city-state was comparatively small, and the mainland provided proportionately

little arable land. Rome was expanding its territories, and was in need of citizens for soldiering and colonization. Hesiod, Plato, Aristotle, and Polybius, among others, discussed various aspects of the population problem. Hesiod, Xenocrates, and Lycurgus were partisans of the one-child family, Himes claims, and as we shall soon see Plato and Aristotle favored a stationary population. See Himes: *Contraception*, p. 79, and M. Moissides: 1932, 'Le Malthusianisme dans l'Antiquite Greque', *Janus* 36, 169–79, cited by Himes.

[13] See below, p. 103.

[14] Therapeutic abortion was practiced by the ancients as is clear from the testimony of Soranus. See below, p. 152.

[15] It is well known that the Spartans practiced eugenics. The newborn was subjected both to (1) the father's right to kill any of his defective offspring as well as (2) the state's right to inspect and kill babies that appeared too weak. According to Durant (*Civilization*, II, p. 81) " . . . any child that appeared defective [to the Spartan council of inspectors] was thrown from a cliff of Mt. Taygetus, to die on the jagged rocks below." Infanticide was also practiced at Athens, though it was not a legal obligation, nor did the state engage in inspection or eugenics *per se*. None of this must obscure the fact that healthy infants were killed throughout Greece and Rome; though public opinion was mixed on the morality of this, it was tolerated.

[16] During Hellenistic times, abortion was punishable by the law only if a husband's consent was not first secured by his wife. One motive for a wife not securing it was her desire to conceal her adultery. Hence it was also punishable if a married woman got an abortion at the instigation of her seducer. See Durant: *Civilization*, II, p. 567. Westermarck states that in Rome abortion " . . . was prohibited by Septimius Severus (c. 145–211 A.D.) and Antoninus, but the prohibition seems to have referred only to those married women who, by procuring abortion, defrauded their husbands of children," *Moral Ideas*, I, p. 415.

[17] The Roman satirist Juvenal (fl. 80 A.D.) speaks to this when he instructs a husband "rejoice; give her the potion . . . for were she to bear the child you might find yourself the father of an Ethiopian." *Satires* V. 141, cited by Durant: *Civilization*, III, p. 364.

[18] The main exception here is the Spartan women, who did receive some military training.

[19] This judgment of a female child's worth was naturally linked to emergent social customs and institutional expectations. The revolutionary character of Plato's reforms for women in Greek society, whereby they are made full-fledged rulers, soldiers, and artisans in his ideally conceived state (*Republic* 445b–466d, *passim*), may be understood in part as a bold rejection of the normally low status accorded women in Classical Greek society. Though he does endorse infanticide on eugenic grounds in his *Republic*, Plato does not favor the preservation of the male child over the female child – as did his countrymen at large. Hence Plato's advocacy of equality between the sexes extends to his policies on infant care and selection.

[20] For a complete discussion of their contributions, see Needham, Joseph: 1959, *A History of Embryology*, 2nd revised ed., Abelard-Schuman, New York, pp. 27–31.

[21] Himes *Contraception*, p. 80. See Aristotle, *History of Animals*, IV. 583a.

[22] Sigerist: *History of Medicine*, II, p. 230. Himes devotes an entire chapter to "The History of the Condom or Sheath." In it he concludes that such devices were known in antiquity, at least by Imperial Roman times " . . . though we know nothing of the extent

of such use. In the absence of more positive knowledge, we must assume that it was not in common use." *Contraception*, Chap. VIII, p. 188.

23 Littré: *Oeuvres*, VII, 415, cited by Himes (trans.): *Contraception*, p. 81. This unusual contraceptive formula is in fact given again in the Hippocratic Collection at Littré: *Oeuvres*, VIII, 171.

24 Lucretius: 1913, *On the Nature of Things*, H. A. I. Munro, (trans.), Bell, London, pp. 165—66, cited by Himes: *Contraception*, p. 83.

25 *Ibid*., p. 166.

26 *Ibid*., p. 167.

27 *Soranus' Gynecology*, Owsei Temkin (trans.), Johns Hopkins Press, Baltimore, 1956, p. 62.

28 *Ibid*., p. 63.

29 See my remarks on this matter in Chap. VI, n. 10, above.

30 Durant: *Civilization*, II, p. 568.

31 *Polybius Histories* XXXVI, 17, W. R. Paton (trans.): cited by Durant, *ibid*., II, p. 568.

32 Durant: *Civilization*, II, p. 567.

33 This definition purposely does not conform in every detail to the latest lexical definition now in use by contemporary physicians. One variant of the latter standardly includes what laymen often call miscarriages. Hence nowadays abortion is defined in medical parlance as "termination of pregnancy, *spontaneously* or by induction, prior to viability. Thereafter termination of pregnancy is called delivery," according to André E. Helleger's article 'Abortion: Medical Aspects', *EB*, I, pp. 2—3. The definition I am using, however, captures the central sense in which the term was usually employed in the ancient medical literature, especially when moral appraisals were offered.

34 Sigerist: *History of Medicine*, II, p. 230. In general, see Hähnel, R.: 1936, 'Der künstliche Abortus in Alterum', *Archiv für Geschichte der Medizin* 29, 224, ff.

35 Littré: *Oeuvres*, VIII, 610, cited by Sigerist, *ibid*., II, p. 230.

36 Sigerist, *ibid*., II, p. 230.

37 Durant: *Civilization*, II, p. 287; and III, p. 365. See also Westermarck: *Moral Ideas*, I, p. 40; and Lecky: *European Morals*, II, p. 28.

38 Terence *Heauton-timorumenos*, Act III, Scene 5, cited by Lecky, *ibid*., II, p. 28, n. 1.

39 *Ibid*., II, p. 28, n. 1. Emphases added. Another similar literary example occurs in Apuleius *Metamorphosis*, X. It too involves a mother's pity for her female infant that her husband commanded her to kill.

40 Durant states: "Sometimes, in the first century, girls or illegitimate children were exposed, usually at the base of the Columna Lactaria — so named because the state provided wet nurses to feed and save the infants found there." *Civilization*, III, p. 364. Even so, it is true that more often than not exposed infants throughout antiquity died.

41 *Collected Dialogues*, F. M. Cornford (trans.), p. 854.

42 Sigerist: *History of Medicine*, II, p. 232.

43 Soranus: *Gynecology*, Temkin (transl), p. xl.

44 See Chap. VI, n. 16, above.

45 Notice that *unlike* the Oath's prohibition against surgery in its very next paragraph (See P5, p. 69, above), in the case of abortion the Oath-taker does not say that he "will withdraw in favor of such men [or women] as are engaged in this work." This strongly implies that the Oath's prohibition against abortion is absolute. Moreover, it is not intended to protect the holiness of the physician *alone*. It is intended to protect the

patient and the fetus from moral wrongdoing as well. Moreover, Durant (*Civilization*, III, p. 313) states that by the time of the Roman Empire, women were engaged in medicine not only as midwives but as doctors. Though he may exaggerate their numbers, he claims: "There were many women physicians; some of them wrote manuals of abortion, which were popular among great ladies and prostitutes."

[46] Prophyry *Pros Garaun*. See Edelstien: 'Hippocratic Oath', p. 19, n. 42.

[47] Edelstein: 'Hippocratic Oath', p. 19. Edelstein also notes that the ascetic rigorism engendered in the Pythagorean way of life, scientific beliefs aside, militated against abortion. For example, extramarital relations were banned and coitus, even in matrimony, was only considered honorable if it aimed at producing children.

[48] Diogenes Laertius *De vitis* VIII. 28–29.

[49] *Ibid.*, VIII. 36, cited by Robinson (trans.), *Early Greek Philosophy*, p. 61. Emphasis added.

[50] Frankena: 'Respect for Life', p. 25.

[51] This observation adds some support to the ancient tradition which ascribes to Pythagoras possible travels to the East, perhaps as far as Babylonia between the time he left Samos and before he settled in Croton, on the east coast of southern Italy, around 530 B.C. But Herodotus is surely mistaken in ascribing to the Egyptians (*Histories* II, 123) a belief in the transmigration of souls, and further implying that the Pythagoreans (though he does not name them) borrowed that doctrine from Egypt. In any case, there is no evidence that the Pythagoreans ascribed souls to vegetative life, or regarded it as sinful to take such life. This is why, strictly speaking, they do not fully qualify as embracing a *comprehensive* respect for life ethic, though they come close.

[52] See Chap. III, above.

[53] At *Republic* X. 621b–c, Er states that after the souls which he observed in the afterworld had suffered their punishments and rewards and had selected their next earthly, soon-to-be embodied lives, each drank from the River of Forgetfulness. He states: "When they had fallen asleep, at midnight there was thunder and an earthquake, and in a moment they were carried up ... *to their birth* ... " Cornford (trans.): *Republic*, p. 359. Emphases added.

[54] *Republic* V. 461. Cornford (trans.), p. 161.

[55] *Ibid.*, p. 161.

[56] The relevant passage on which I base this claim is *Republic* V. 461a–c.

[57] For a somewhat less restrictive and seemingly more humane set of regulations bearing on matters relating to population control and the abortion option, see Plato's *Laws* V. 740d. There, Plato does not say that he endorses abortion or infanticide in order to control populations in excess of 5,040 – a number he deemed ideal for the state. In fact, he refers to setting up colonies abroad from the mother country as "the last resort." Hence, it is conceivable that in his old age Plato drew back a bit from the position he advocates at *Republic* V. 462. Plato's handling of adulterous citizens is also greatly toned down at *Laws* VI. 784–5 and VIII. 841, ff.

[58] *Republic* V. 460, ff. Cornford (trans.), p. 160.

[59] Aristotle *Politics* VII. 1265b 10, ff. Barker, Ernest (trans.): 1946, *The Politics of Aristotle*, Oxford University Press, Oxford.

[60] *Politics* VII. 1265b 5, ff.

[61] See p. 104 above. The ideal size of a city-state, according to Aristotle, was 10,000.

62 Barker (trans.): *Politics*. Emphases added. The Greek of the final sentence reads, in part: " . . . *tei aisthesei kai to zen estai*."

63 See Aristotle's *Nicomachean Ethics* I. 8; and *Rhetoric* I. 5.

64 See Chap. III, p. 50, above.

65 Another word in Greek meaning "life" was *bios*. This word referred not to biological life *per se*, but rather to an individual's personal experiences as in "the course of life." Liddell and Scott: *Greek Lexicon*, p. 130. The word *zoa* in Greek also meant life, but especially life in the sense of what we would now call biological subsistance (*ibid*., p. 297). Hence when to this list is added a third Greek word, *psyche*, meaning in our day, the "life of the spirit" or "mind," but capable of meaning in Aristotle's day life *simpliciter*, it becomes imperative that the concept of life Aristotle intended in his writings on abortion be specified. Minimally, he means that the human fetus ought not be aborted if it possesses life (*zen*) at such a level that it has the sensitive capacity (*aisthesis*) to feel pleasure or pain. This latter capacity is found in almost all animals and in higher forms of life, according to Aristotle.

66 Significantly, the verb *empsychoo* in Greek meant "to animate." The Latin word *anima* was associated with the Greek word for soul (*psyche*), and was often used to contrast the body. Also, *anima* originally denoted the sign of life: breath. (Liddell and Scott, *ibid*., p. 798.) It is directly from this Latin word *anima*, not the Greek *psyche*, that we get the animate-inanimate distinction in English. This is a distinction standardly made between things that are alive and capable of self-movement, and those that are not. Aristotle, moreover, held that plants moved in the minimal sense of undergoing generation, growth, and decay. Plants did not, however, possess locomotion, as did most (but not all) animals.

67 Ross, David: 1974, *Aristotle*, Harper and Row, New York, p. 134. See also Ross, pp. 129–42, for a more detailed account of Aristotle's psychology.

68 The relevant passages are: *De Anima* 414a 2–4; 415a 3–6; 435a 12. Aristotle has it that as an outgrowth of the sensitive soul there occurs in most animals, on the cognitive side, imagination (of which memory is a further development); and, on the appetitive side, the capacity for movement (locomotion).

69 Aristotle elsewhere states: "What is called *effluxion* is a destruction of the embryo within the first week, while *abortion* occurs up to the fortieth day." *Historia Animalium* VII. 3. 583a 10, ff. Emphases added. His switch from 'effluxion' to 'abortion' in this passage cannot have been arbitrary, it seems to me. "Effluxion" probably signals the conceptus possessing the nutritive soul alone. See *The Works of Aristotle*, eds., by J. A. Smith and W. D. Ross, D. W. Thompson (trans.), 11 vols., Clarendon Press, Oxford, 1910, IV, p. 583.

70 *Historia Animalium* VII. 3. 583b, Thompson (trans.): *Ibid*., p. 583. By "distinct parts," Aristotle means the presence of heterogeneous organs such as hands, eyes, heart, etc.; by "fleshlike substance," he means homogeneous constituents such as tissue, blood and bones. In his embryological theory, Aristotle embraced a theory of epigenesis according to which the individual organism develops by structural elaboration of the unstructured zygote rather than by a simple enlarging of a preformed entity.

71 According to Crawley, A. E.: 'Foeticide', *ERE*, VI, p. 56, Aristotle's followers grappled with this issue, attempting to bring certainty and consistency to his teachings on abortion. "His followers distinguished between the male and female embryo in the

date of animation. The male was regarded as being forty days after conception, the female eighty days [sic]. Later the moment of animation was fixed for both sexes at the fortieth day. The Roman jurists adopted the latter view. The general distinction between the animate and the inanimate foetus was clearly held by Canon and Roman law alike, and lasted to modern times. It was applied by Augustine thus: 'The body is created before the soul. The embryo before it is endowed with a soul and is *informatus*, as its destruction by human agency is to be punished with a fine. The embryo *formatus* is endowed with a soul; it is an animate being: its destruction is murder, and is to be punished with death.' " Augustine *Questiones in Exodum* LXXX; *Questiones Veteris et Novi Testamenti* XXIII, cited by Crawley. Variants of this distinction endured in Roman Catholic theology until A.D. 1869, at which time Pope Pius IX formally eliminated the distinction between an animated and unanimated fetus. He further decreed that all fetuses should be considered ensouled from conception and that therefore *any* abortion was equivalent to murder.

72   Seneca *Consolation to Marcia* XXIV. 4. M. Hadas (trans.), cited by Choron: *Death and Western Thought*, p. 69. Emphases added.

73   Seneca *Consolation to Marcia* XXII. 3.

74   As a major Hellenistic philosophical movement, Stoicism flourished for well over five centuries. It originated in the fourth century B.C. with its founder, Zeno of Citium (c. 336–264 B.C.) and counted among its many adherents the Roman Emperor Marcus Aurelius who ruled from 161 to 180 A.D. Traditionally, Stoic thought is divided into three main periods during which there occurred significant changes in doctrine. Roughly, these periods are early (336–186 B.C.); middle (185–5 B.C.); and late (4 B.C.–200 A.D.). Seneca belongs to the last of these periods. Yet he was influenced by the Middle Stoics Panaetius and Posidonius, with whose views on the development of the human soul I here associate him. Seneca wrote no treatise on human embryology. My account of his theory of human personhood is thus necessarily speculative. I am indebted to Michael Frede, of Princeton University, for clarifying some of the details of Stoic moral psychology. An unusually concise and useful treatment of the Stoics can be found in Sandbach, F. H.: 1975, *The Stoics*, W. W. Norton, New York.

75   *Stoicorum Veterum Fragmenta* II, cited by Edelstein: 'Hippocratic Oath', p. 18, n. 42. For a useful description and analysis of Stoic psychology, see Hicks, R. D.: 1962, *Stoic and Epicurean*, Russell and Russell, New York, pp. 60–3. Concurrent with these Stoic doctrines, it may be observed that Roman law and custom distinguished sharply between abortion and infanticide in the following way. The unborn child was not said to be *homo* (a human being), nor was it even an *infans* (infant). Rather it was said to be a *spes animantis* ("the prospect of a living being"). See Crawley: 'Foeticide', *ERE*, VI, p. 56.

76   This useful distinction between the *moral* and *biological* (or what she calls the "genetic") sense of the term "human" has been pointed out by Mary Anne Warren. See her 'On the Moral and Legal Status of Abortion', *The Monist* 57, (1973), pp. 43–61. In general, unless I indicate otherwise or the context clearly demands the biological sense of "human" or "person," it is personhood in the *moral* sense with which this essay is most concerned and to which it usually refers. For this is the crucial sense on which most bioethical theories and decisions of the sort examined here come ultimately to rest.

77   See above, Chap. VI, n. 7, and Crawley, *ibid.*, VI, p. 56, who has it that the father *alone* could legally order an abortion in Rome.

[78] Spangenberg: 'Verbrechen der Abtreibung der Leibesfrucht', in *Neue Archiv des Criminalrechts*, II, 22, ff., cited by Crawley, *ibid.*, VI, p. 56 and Westermarck: *Moral Ideas*, I, p. 415. The modern reader recognizes in this Stoic defense of the discretionary right of a woman to seek an abortion, the paradigm of contemporary arguments of a similar sort. In particular, there is the contemporary, pro-abortion argument that asserts a woman's putative property rights over her body, likens the fetus to any one of her organs, and concludes that the discretionary right of a woman to have her appendix removed extends, by analogy, to her right to abort or not. Criticisms against these and other contemporary variants of Stoic and related ancient positions for and against abortion are readily available. For specific criticisms of the property rights argument, see Margolis, Joseph: 1976, *Negativities: The Limits of Life*, Charles E. Merrill Publishing Co., Columbus, Ohio, Chap. III, p. 45, ff., and Feinberg, Joel: 1980, *Matters of Life and Death*, ed. by Tom Regan, Random House, New York, pp. 203–05.

[79] Seneca *De ira* I. 15.

[80] Hicks *Stoic and Epicurean*, p. 26.

[81] Some commentators are fond of quoting Seneca's *Consolation to Helvia*, his mother, in which (at XVI. 3) he praises her for not crushing " . . . the hope of children that were being nurtured in your body," as if Seneca were opposed to abortion. This, however, misses the point. The Stoics never argued that women ought to abort, only that it was their *discretionary right* to do so. In other words, it is up to the woman, at her discretion alone, whether to abort or not: she has neither the strict duty to bear, nor the obligation to abort her child. Hence, although her choice is sovereign in respect to the fate of the fetus, this does not bar the moral community from holding the conduct of some women more thoughtless or frivolous than that of others in their exercise of this right. It is in this tradition that Seneca praises his mother. For he personally holds in contempt women whose " . . . only recommendation lies in their beauty," and who therefore regard pregnancy " . . . as if an unseemly burden," as he says in that same passage. See *Seneca: Moral Essays*, John W. Basore, (trans.), William Heinemann, London, 1932, II, pp. 471–73.

CHAPTER VII: THE PROBLEM OF EUTHANASIA

[1] For example, euthanasia is now also standardly qualified as *involuntary* euthanasia, evidenced by the absence of the patient's consent to quit life, and *active* versus *passive* euthanasia, discussed below, p. 133. Sissela Bok provides a straightforward account of euthanasia in its various modern senses in 'Death and Dying: Euthanasia and Sustaining Life', *EB*, I, pp. 268–77.

[2] See Chap. III, pp. 47–48, above, particularly points (2) and (3).

[3] Philo attributed this view to the Hellenistic masses (*"hoi polloi"*). Philo *Vit. Mos.* I, 11, cited by Jakobovits, Immanuel: 1976, *Jewish Medical Ethics*, Bloch Publishing Co., New York, p. 174, n. 51.

[4] This shift in emphasis is supported by examples of more recent uses of the term in the last century or so, as recorded by the *Oxford English Dictionary*, 12 vols., Clarendon Press, Oxford, 1961, III, p. 325, in regard to the English word euthanasia.

[5] Daube, David: 1972, 'The Linguistics of Suicide', *Philosophy and Public Affairs* 1, 391.

[6] *Ibid.*, p. 391.

[7] See Rudolf Hirzel's classic 'Der Selbstmord', *Archiv für Religionswissenschaft* 11, (1980), p. 243, for which I thank Truman Eddy for his kind assistance. See also Aristophanes' *The Frogs* 116, ff., wherein he alludes to the advantages of hemlock over other methods as an aid to suicide; and Theophrastus *History of Plants* IX. 16. 8, wherein he remarks on the powerful and incurable effects of poisons produced from hemlock and poppy.

[8] Daube: 'Linguistics of Suicide', p. 393. Daube gives Diodorus Siculus 2. 57. 5, as one ancient source of the myth to which his quote refers.

[9] See Margolis' *Negativities*, Chap. II, 'Suicide', pp. 23–35, which is also reprinted in Beauchamp, Tom L. and Perlin, Seymour: 1978, *Ethical Issues in Death and Dying*, Prentice Hall, New Jersey, pp. 92–7.

[10] Daube: 'Linguistics of Suicide', pp. 399–400; 406. Daube gives as sources for these Greek expressions: Xenophon *Hellenica* VI. 2. 36; Euripides *Iphigenia in Tauris* 974; Plato *Phaedo* 68a; Aristotle *Nicomachean Ethics* IX. 4. 8. 1166b; Diogenes Laertius *Lives* VII. 130 (on Zeno).

[11] *Ibid.*, p. 403. Daube (p. 397) offers the anthropological theory that, regarding the evolution of any native language in a given culture, "The emergence of the noun betokens a decisive advance in abstraction, systematization, institutionalization." If so, the regular appearance of Greek nouns to refer to suicide suggests a trend toward wider acceptance or at least reflection concerning the practice.

[12] I have benefited from the literary and historical evidence amassed in Mair, A. W.: 'Suicide', *ERE*, XII, pp. 26–33.

[13] The prevalent social attitudes of antiquity, however, tended to construe suicidal behavior, which nowadays is widely classified as symptomatic of a medical problem, as alternatively symptomatic of moral problems or general problems in living. This observation suggests the normative character of the ascription of illness, particularly mental illness – a topic I shall not treat here. For more on mental disorders in antiquity, see Rosen, George: 1971, 'History in the Study of Suicide', *Psychological Medicine* 1, 272. In general, see Margolis: *Negativities*, Chaps. VII–VIII.

[14] This statement stands despite my remarks in preceding n. 13. For the ancients did in fact recognize as involuntary and hence blameless *some* instances of supposed mental disorder. But they counted such instances far fewer in number. See Hippocrates *Epidemics* VII. 89, cited by Rosen: 'Suicide', p. 272, as one example.

[15] Homer *Odyssey* XI. 548; and Sophocles *Ajax*.

[16] Lycurgus *c. Leokrat.* 84, ff., cited by Mair: 'Suicide', *ERE*, XII, p. 27.

[17] Plutarch *Thes.* XXII.

[18] *Herodotus* I. 45.

[19] Theogonis 173, ff., cited by Mair: 'Suicide', *ERE*, XII, p. 27.

[20] Ovid *Her.* V, cited by Mair, *ibid.*, XII, p. 28.

[21] Plutarch *Mulierum Virtutes* 249b–4, cited by Mair, *ibid.*, XII, p. 29.

[22] Strabo X. 486, cited by Mair, *ibid.*, XII, p. 29.

[23] Pliny (the younger) *Epistle* I. 22. See also Diogenes Laertius VII. 130; and Pliny *Epistle* III. 7 for related illustrations.

[24] Mair: 'Suicide', *ERE*, XII, pp. 30–1; 34.

[25] The first recorded theoretical discussion of *active euthanasia* occurred in Thomas More's *Utopia*, first published in 1516, although the term "euthanasia" had not yet been

explicitly coined. In its modern sense, the term first appeared around 1870 according to Fye, W. Bruce: 1979, 'Active Euthanasia: An Historical Survey of Its Conceptual Origins and Introduction Into Medical Thought', *Bulletin of the History of Medicine* 52, pp. 492–94, *passim*.

26 *Phaedo* 62b–c, Hugh Tredennick, (trans.), *Collected Dialogues*, p. 45. Emphases added. In an interesting article ('Who *Did* Forbid Suicide at *Phaedo* 62B?', *Classical Quarterly* XX, 1970, 216–220), J. C. G. Strachan suggests that – contrary to J. Burnet and R. Hackforth, among others – this allegory, its body-guardpost imagery, and its accompanying rationale opposing suicide, may not have belonged either to Philolaus (as was assumed even in later antiquity) or to the fifth century B.C. Pythagoreans of Plato's day. The thrust of his argument, though far from conclusive, turns on (a) emphasizing the hiatus and possible ambiguity in the text between the introduction of Philolaus' name at 61d ff., and the introduction of the allegory at 62b which bears no explicit attribution to Philolaus; (b) establishing the far firmer association of the allegory at 62b with Orphic, not Pythagorean, religious teachings by the fifth century B.C. period in question; and (c) underscoring the subsequent confusion of sorting out the true owner-ship of the allegory at 62b, which was occasioned by the eventual acceptance of the body-prison doctrine by the Pythagoreans toward the end of the fourth century B.C. (a confusion which also allegedly misled Circero in his *De Senectute* 20, and modern scholars who followed). Strachan has, I think, succeeded in undermining any *unqualified* ascription of this passage to Philolaus. But he is surely fanciful in suggesting that perhaps " . . . Plato was unwittingly responsible for the adoption of this particular piece of Orphism by the Pythagorean school" (p. 220). For more on this controversy, see Burnet, J.: 1911, *Plato's Phaedo*, Clarendon Press, Oxford, pp. 22, ff.; Hackforth, R.: 1960, *Plato's Phaedo*, Bobbs-Merrill, Indianapolis, Indiana, p. 38; and Bluck, R. S.: *Plato's Phaedo*, Bobbs-Merrill, Indianapolis, Indiana, p. 195, ff., cited by Strachan.

27 See Chap. VI, n. 65, above.

28 Frankena: 'Respect for Life', pp. 25–6.

29 *Ibid.*, p. 26.

30 *Ibid.*, pp. 38–42.

31 Tredennick, Hugh (trans.): *Collected Dialogues*, p. 25. Emphases added.

32 *Phaedo* 62c–4. Jowett, Benjamin (trans.): *Republic and Other Works*, Doubleday, New York, Doubleday, n. d.

33 See Chap. VI, pp. 114–115, above.

34 *Republic* III. 407d. Cornford, Francis (trans.): 1973, *Republic of Plato*, Oxford University Press, Oxford, p. 97. Emphases added.

35 Such, in fact, is the precise picture we get of the infamous gymnastic master and trainer, Herodicus. Socrates describes Herodicus in unflattering terms. He is one who, after losing his health " . . . combined training and doctoring in such a way as to become a plague to himself first and foremost and to many others after him . . . by lingering out his death. He had a mortal disease, and he spent all his life at its beck and call, with no hope of cure . . . . Every departure from his fixed regimen was a torment; and his skill only enabled him to reach old age in a prolonged death struggle." *Republic* III. 406b–c. *Ibid.*, p. 96. Cf. Heidel, William A.: 1941, *Hippocratic Medicine: Its Spirit and Method*, Columbia University Press, New York, p. 123.

36 See Mair: 'Suicide', *ERE*, XII, pp. 30–1; and Rose, Herbert J.: 'Disposal of the Dead', *OCD*, pp. 314–15.

[37] *Laws* 873c–e. Saunders, Trevor J. (trans.): 1976, *The Laws*, Penguin Books, New York, p. 391. Note that this fourth condition is not, as qualified in the text, equivalent to what I classify as a form of heroic suicide. For to be heroically characterized, the subject and others must regard his disgraceful deed as genuinely accidental and unintended. In contrast, the fourth condition in *Laws* simply excuses the subject who has "... fallen into some irremediable disgrace he cannot live with." The tragic element of not intending to cause such disgrace is missing, or, at least, not required by Plato. Cf. pp. 132–133, above.

[38] *Laws*, Saunders, (trans.), p. 391.

[39] *Ibid.*, p. 391.

[40] Edelstein contends that under the onslaught of the later Stoic attack on their views, sometime after Aristotle died, his disciples deserted his teachings against self-murder. 'Hippocratic Oath', p. 17 and n. 36.

[41] *Nicomachean Ethics* V. 10. Ross, W. D. (trans.): *Basic Works*, pp. 1020–21.

[42] For Aristotle's full quotation on death and courage, see Chap. III, pp. 50–51, above.

[43] *Nicomachean Ethics* III, 1115b 7. Ross (trans.): *Basic Works*, p. 975.

[44] *Ibid.*, 1115b 18, p. 976.

[45] *Ibid.*, 1116a 12–15, p. 977. Emphases added.

[46] Edelstein: 'Hippocratic Oath', p. 17. Edelstein tries very hard to downplay Aristotle's moral censure of suicide. If so, was this done in order to protect his own theory concerning the exclusive Pythagorean origins of the Hippocratic Oath? See Chap. V, above.

[47] This distinction is discussed on pp. 135–136, above.

[48] See pp. 120–121, above. There is some evidence that the late Stoic Seneca occasionally flirted with a version of *divine personal immortality*. But he admits that on this score philosophers of old (referring to Plato) promised more than they could deliver, and doesn't seem to take the notion of personal immortality seriously. Such a notion is not at any rate central to Stoic teaching, and I am convinced that Choron (*Death and Western Thought*, p. 69) exaggerates Seneca's dependence on this perspective. Cf. Seneca *Epistles* XXIV and CII.

[49] The metaphor of life as a banquet, at which only the greedy guest overstays his visit, was a favorite of Seneca's.

[50] The Cynics based the right to suicide on the prior right to be able to separate oneself from one's community at any time. They advocated an *unlimited* right to commit suicide, therefore, which the Stoics refused to grant. See Hirzel: 'Der Selbstmord', p. 279, ff. As for the Cyrenaics, it is true that *only some* of them endorsed suicide on moral grounds. Theodoros opposed it unconditionally. But Hegesias argued the case for rational suicide so eloquently that a rash of suicides occurred in Alexandria, and Ptolemy II ordered him banished from Egypt in consequence. See Diogenes Laertius, *Lives* II, 86; and Cicero *Tusc.* I. 34. The Epicureans were officially opposed to suicide since the wise man neither seeks death as a refuge nor declines life – fearing nothing. Even so, the Roman Lucretius broke with Epicurus on this issue. He granted suicide limited warrant for those to whom life had grown distasteful. It appears, moreover, that many late Epicureans sided with Lucretius, who himself is reported to have committed suicide. For more on the Epicureans and suicide, see Choron, Jacques: 1972, *Suicide*, Charles Scribner's Sons, New York, pp. 112–15.

[51] On death as a test of moral strength, see Seneca *Epistle* LXX. 17–19.

[52] Tacitus (*Annals* W. 60–63) sympathetically records the manner of Seneca's death,

which showed much bravery. It involved the initial supply of a poison by Seneca's physician, which did not act as speedily as hoped for.

53 Seneca *Epistle* LXX. 14–18. *Seneca: Ad Lucilium Epistulae Morales*, Gummere, Richard M. (trans.): 1962, 3 vols., Harvard University Press, Cambridge, Massachusetts, II, p. 65. Emphases added. Critics of the Stoic position in support of suicide have long observed an inherent contradiction in the Stoic perspective. On the one hand, Stoics claim that the wise person must resign himself to the course of destiny by making his will conform to what must happen anyway. On the other hand, they allow that when a person is placed in circumstances calling for intolerable or unworthy endurance, this resignation can justifiably give place to active resistance. As E. Zeller has challenged: Is it " . . . consistent with unconditional resignation to the course of the world, to evade by personal interposition, what destiny with its unalterable laws has decreed for us?'' There appears to be no clear way out of this dilemma for the Stoics, though I must waive that controversy here. See Zeller, E.: 1962, *The Stoics, Epicureans and Sceptics*, Russell and Russell, Inc., New York, p. 339.

54 Seneca *Epistle* LXX.

55 Zeller: *Stoics*, p. 337. See *Epistles* LVIII. 36; and LXX. 11.

56 Zeller: *Stoics*, p. 336.

57 Seneca *Epistle* XXIV. 24–26. Gummere (trans.): *Seneca*, I, p. 181.

58 The first claim is critically discussed by Zeller: *Stoics*, pp. 338–39.

59 Seneca *Epistles* LXV and CII. See also Chap. VII, n. 48, above.

60 Mair: 'Suicide', *ERE*, XII, p. 31. The passage in question from *Phaedo* 62C was also treated on pp. 137–138, above. Mair credits the later Stoics Panaitios (c. 140 B.C.) and Posidonios (c. 130–46 B.C.) for modifying Socrates' emphasis on *external* compulsions which force suicide, to *internal* compulsions such as inner, overmastering impulses of ambiguous origin. Cf. Seneca *Epistle* CII. 23. In any case, note that one result of such a reinterpretation meant that a Stoic could be said to have chosen suicide on religious grounds, if he took his internal impulses to end life as the divine decree of heaven.

61 Seneca himself wrote that he once considered ending his life as a young man on account of an annoying illness (catarrh) that caused him to become dangerously thin. But discovering that his prognosis was not terminal, he endured and overcame his sickness with the help, he says, of philosophy and friends. He gives as his primary reason for not killing himself at the time his profound love for his father, and the grief his suicide would cause the old man. See Seneca's *Epistle* LXXVIII.

62 The question, however, can be raised as to whether Aristotle could be plausibly said to be a proponent of the respect for human life ethic at least *prima facie*, in W. D. Ross' sense. This is certainly conceivable, though one will search in vain for textual evidence in Aristotle to satisfy such an approximate reinterpretation. Even so, the point remains that the concept of respect for human life admits of at least two possible forms: (1) the absolute form, which states that it is *always* actually wrong to end or prevent a human life; or (2) the non-absolute form, which states that it is *not always* or finally wrong to prevent or shorten a human life, since other overriding considerations may still make it right in certain instances. I have focused on the first form as this is the most useful expression of the principle in relation to my analysis of the origins of the Hippocratic Oath, particularly its absolutely formulated prohibitions against abortion and euthanasia. See Ross, W. D.: 1930, *The Right and the Good*, Clarendon Press, Oxford, p. 19, ff.; and Frankena: 'Respect for Life', pp. 32–33.

CHAPTER VIII: THE PHYSICIAN'S MORAL RESPONSIBILITY

[1] Hart, H. L. A.: 1968, *Punishment and Responsibility*, Oxford University Press, Oxford, pp. 211–30. Such role responsibilities can be classified into two types of duties: (a) moral duties and (2) legal duties. I shall mainly deal with the former.

[2] Hippocrates: 1952, *On Intercourse and Pregnancy*, Ellinger, Tage (trans.), Henry Schuman, Inc., New York, p. 48; and Littré: *Oeuvres*, XIII, 7, p. 490.

[3] Soranus: *Gynecology*, Temkin (trans.), p. 63.

[4] See Chap. VI, pp. 105–106, above.

[5] Largus, Scribonius: 1887, *Compositiones*, ed. by Georg Helmreich, Teubner, Leipzig, p. 2. Scribonius and Erotian took Hippocrates to be the true author of the Oath.

[6] *Ibid.*, p. 2.

[7] *Ibid.*, p. 2. Scribonius Largus' concept of humaneness (*humanitas*), which was a central idea in his medical philosophy, meant more than simply affecting a friendly disposition. It involved what he called a " . . . proficiency and benevolence toward all men without distinction." It is safe to say that this amounted for him to a love of mankind without regard to race, creed, or color. Though there is no evidence that he was a Christian convert, he may have been tinged by Christian or Stoic notions of the universal brotherhood of man. At any rate, Scribonius was aware that his philosophy of medical humanism went considerably beyond the moral scope of the Hippocratic writings on ethics and etiquette.

His use of the word *humanitas* corresponded, at the time he wrote, with the then dominant meaning of the Greek word *philanthropia* (the love of mankind). Yet it is not likely that the Hippocratic Oath was consciously animated by *philanthropia* at the time it was written; that sentiment was not native to Greek thinking in pagan times. Even so, I must concur with Edelstein's judgment that the Oath nonetheless " . . . assumes full significance and dignity [from the modern point of view] only if interpreted in the way in which it was understood by Scribonius and those who came after him." Edelstein: 'The Professional Ethics of the Greek Physician', *Ancient Medicine*, p. 374.

[8] Temkin, Oswei: 1975, 'The Idea of Respect for Life in the History of Medicine', in *Respect For Life: In Medicine, Philosophy, and the Law*, ed. by Stephen Barker, Johns Hopkins Press, Baltimore, pp. 10–11.

[9] But to what degree did pagan followers of the Oath feel this conflict of duty? If they were devoutly religious, as presumably many were, is it not likely that they saw the irresistible will of heaven in the tragic loss of the mother's life during childbirth? Indeed, have not some modern partisans of the Hippocratic Oath's stand on abortion traditionally found solace in a similar theological conviction, i.e., that the regrettable loss of the mother, or fetus, or both, is God's will?

It will be recalled that the ancients did not construe the question of abortion (therapeutic or otherwise) in the now familiar language of competing moral rights. For example, weighing the putative right to life of the fetus versus that of the mother, community, etc. The claim that moral rights exist is mainly a modern one.

[10] The Caesarean section is one of the oldest surgical procedures. It was almost always fatal to women throughout antiquity and the Middle Ages. Legend had it that Julius Caesar was born in this fashion, which is most doubtful. The operation is named for a branch of the ancient Roman family of the Julii, whose cognomen, Caesar (cf. cadere, "to cut"), originated from a birth by this means, according to Pliny. 'Caesarean Section',

Encyclopaedia Britannica Encyclopaedia Britannica Inc., London, 1971, IV, p. 577. Zeus performed a version of the Caesarean when he rescued his still unborn son, the infant Dionysus, from the womb of his mortal lover, Semele, who had suffered a tragic and sudden death. Fortunately for mankind, Dionysus flourished.

[11] One religious consideration related to the Pre-Christian acceptability of physicians supplying abortifacients was this. By Hellenistic times, it was not widely believed that the ghosts of terminated fetuses sought revenge against their living offenders, if indeed that was ever a serious constraint. There is evidence suggesting that, in earlier times, some types of ghosts were thought to "walk" or to be restless, but few sought revenge against the living for their misery. It is true that Plato records the souls of some suicides suffering penance in the afterworld (*Republic* X. 615c). But the absence in pagan thought of anything like the Christian doctrine of Original Sin, which eventually was thought to damn the souls of fetuses whose lives were ended without baptism, must have lessened the religious anxieties of the pagans in regard to abortion. See Mair: 'Suicide', *ERE*, XII, pp. 30—31.

[12] *The Art* III, Jones (trans.): Hippocrates, II, p. 193.

[13] *Prognostic* I, Jones (trans.); *Hippocrates*, II, pp. 7—8. Other relevant passages in the Hippocratic Collection which show a marked concern for professional reputation include *Precepts* IX; and *Decorum* XI, XVII.

[14] Hippocrates *Pyorrhetic* II, Littré: *Oeuvres*, VI, pp. 4—6. See Gourevitch, Danielle: 1969, 'Suicide Among the Sick in Classical Antiquity', *Bulletin of the History of Medicine* 43, 501—18. One must beware of inferring from this discussion that the Hippocratics regularly shunned difficult cases. For evidence to the contrary, see *On Joints* LXIX; *Ancient Medicine* IX; and *Precepts* VII.

[15] In an interesting paper, Edelstein has argued that the role of religious or temple medicine for the terminally ill, by the fifth century B.C. and for centuries later, was roughly this. Those patients declared untreatable often sought refuge and alternative therapies of a supernatural sort under the auspices of the temple priests of Asclepius. The Hippocratic physicians did not customarily prescribe such religious therapies themselves, it is true; but they studiously left that option open. Hence, those of their patients declared untreatable were not in fact deserted by all representatives of the medical community: the religious segment was a last resort, providing comfort and solace, if not lasting cures, to the suffering multitude. Edelstein: 'Greek Medicine in Its Relation to Religion and Magic', *Ancient Medicine*, pp. 205—46.

[16] Could another factor in the physician's motivation have been his financial gain through the sale of lethal drugs? This possibility cannot be dismissed. As suggested in Chap. I, most physicians had to work for a living; some had trouble making ends meet. But as Apuleius' *Metamorphoses* implies, it is entirely unwarranted and even slanderous to depict the ancient physician as an unseemly poison-peddler who practices only for the sake of his purse. For more on Apuleius, see Chap. V, n. 46, above. Plato's ideal physician functions *qua* physician to serve others, and only in a subsidiary capacity does he function as a wage-earner (*Republic* I. 341c, ff.). In general, see Kudlien, F.: 1976, 'Medicine as a "Liberal Art" and the Question of the Physician's Income', *Journal of the History of Medicine* 31, No. 4, pp. 448—59.

[17] Gourevitch: 'Suicide Among the Sick', pp. 508, ff.

[18] The passage about to be discussed (construed *primum non nocere* in Latin texts) is also examined in Chap. V., p. 76 and p. 83, above, in a somewhat different context.

[19] *Epidemics* I.XI. Jones (trans.): *Hippocrates*, I, p. 165. Emphases added.

[20] Abortion was formally condemned in the West by the Church as early as 305 A.D. by the Council of Elvira meeting on the Iberian Peninsula (canon 52); later it was condemned in the East within a decade by the Council of Ancyra in 314 A.D. (canon 21), according to Noonan, J. T.: 'An Almost Absolute Value', in *Morality of Abortion*, p. 14. Severe penalties were imposed, including excommunication for women who aborted in order to conceal adultery. For further developments during the early Christian Period, see Noonan, pp. 11–18 *passim*. As for euthanasia, Amundsen has observed ('The Physician's Obligation to Prolong Life', p. 27) that while the early Christian tradition held it to be a sin for the physician to terminate life using active euthanasia, " . . . it laid no stress, apparently, on the positive correlate that would require the physician actively to prolong life." The Christian position was based largely on the principle that God alone gives life and man ought not interfere with God's purposes. Many early Christian writers held that either God caused some persons to fall ill or He did nothing to prevent such diseases from occurring. Secular medicine was thus widely viewed as working against God's divine plan for man and it took many centuries for the medical art to gain acceptance among the devout. Citing Arnobius (*Adversus gentes* I. 48) and Tatian (*Oratio Graecos* 18), Amundsen concludes that "while abortion, suicide, and enthanasia became sins, the prolonging of life did not become either a virtue or a duty," p. 27.

[21] Jones, W. H. S. (trans.), *The Doctor's Oath*, p. 55. For evidence on the possible influence of the Hippocratic Oath on early Jewish medical writers, see Tosner, Fred and Muntner, Sussman: 1965, 'The Oath of Asaph', *Annals of Internal Medicine* 63, 317–20.

CHAPTER IX: CONCLUSION

[1] See, for example, Kristeller, Paul O.: 1961, *Renaissance Thought: The Classic, Scholastic and Humanistic Strains*, rev. ed., Harper and Row, New York.

[2] Hume, David: *An Enquiry Concerning Human Understanding*, Section IV, 1–2; Section V.

[3] Kudlien: 'Medical Ethics and Popular Ethics', p. 109, ff.

[4] Ross, *The Right and the Good*, Chap. II, is instructive here.

CHAPTER X: EPILOGUE

[1] There are two interesting articles by Amundsen that bear on ancient medicine and law: (1) 'The Liability of the Physician in Classical Greek Legal Theory and Practice', *Journal of the History of Medicine and Allied Sciences* 32, April, 1977, pp. 172–203; and (2) 'The Liability of the Physician in Roman Law', *International Symposium on Society, Medicine and Law*, Elsevier, Amsterdam, 1973.

[2] Edelstein: 'The History of Anatomy in Antiquity', *Ancient Medicine*, pp. 247–301.

[3] See Ferngren, Gary: 1982, 'A Roman Declamation on Vivisection', *Transactions and Studies of the College of Physicians* IV, 272–290, which also contains some useful bibliographical sources.

[4] Rosen, George: 1968, *Madness in Society: Chapters in the Historical Sociology of Mental Illness*, University of Chicago Press, Chicago.

[5] McKinney, Loren C.: 1952, 'Medical Ethics and Etiquette in the Early Middle Ages: The Persistence of Hippocratic Ideals', *Bulletin of the History of Medicine* 26, 1–31.

⁶ One looks in vain for the parallel notion of the rights of persons, whether in the role of patients or otherwise, in the ancient moral or legal literature. The whole concept of rights, especially when construed to entail accompanying duties owed by others to the bearer of those rights, is a rather recent moral and legal development originating in 17th century European thought. Nonetheless, it would be mistaken to suppose that the ancient physician was blatantly insensitive to the need of securing his patient's consent before starting treatment. The looming medical principle was that Nature was the master healer. Most physicians realized that for psychological, if not moral reasons, the desired cure was most likely to result if the patient were made to feel a cooperative partner who assisted Nature and the physician by following the prescribed therapy.

⁷ Hippocrates *Decorum* XVI.

⁸ Vaux, Kenneth: 1974, *Biomedical Ethics: Morality for a New Medicine*, Harper and Row, New York, p. 9.

⁹ Cited by Sigerist: *History of Medicine*, II, p. 238.

¹⁰ Hippocrates *Precepts* VIII.

¹¹ Hippocrates *Decorum* XVII.

¹² Altman, Lawrence K.: 1979, 'The Doctor's World', *New York Times*, May 1, p. CL.

¹³ Actually, a case could be made that some free medical attention to the needy and hopelessly ill was dispensed by priest-physicians in Pre-Christian times at the Temples of Asclepius at locations like Epidaurus on mainland Greece. There were 400 Asclepian temples by Aristotle's time. But such free care was at any rate the exception rather than the rule. It is also arguable whether what the patient usually received was medical treatment at all or rather superstitious ritual therapies epitomized by the famed dream-healing process or incubation sleep (judged by Hippocratic standards, at least). It remains true, however, that charity was not a firmly embedded pagan value.

¹⁴ It was not until 390 A.D. that Fabiola, a wealthy Roman widow, founded the first Christian hospital, and as late as 431 A.D. Saint Augustine regarded them as a novelty. Alan T. Marty, who has argued convincingly that public hospitals were *not* uniquely Christian institutions in antiquity, points out that they spread more rapidly in the Eastern Byzantine world "because of the energy and scholarship of the disciples of Nestorius [the fifth century Patriarch of Constantinople] . . . By the middle of the sixth century, however," Marty states, "hospitals were more securely established and their property duly safeguarded" (p. 1022). See Marty: 1971, 'The Pagan Roots of Western Hospitals', *Journal of Surgery, Gynecology, and Obstetrics* 133, 1019–1022 *passim.*

¹⁵ Even so, *Precepts VI* furnishes us with an inviting glimpse of the Hippocratic physician at his best. The author writes: "I urge you not to be too unkind, but to consider carefully your patient's superabundance or means. Sometimes give your services for nothing, calling to mind a previous benefaction or present satisfaction. And if there be an opportunity of serving one who is a stranger in financial straits, give full assistance to all such. For where there is love of man, there is love of the art." Jones (trans.): *Hippocrates*, I, p. 319. See also Chapter V, pp. 91–92, above.

¹⁶ Veatch, Robert M.: 'Codes of Medical Ethics: Ethical Analysis', *EB*, I, p. 177.

¹⁷ The effectiveness of professional codes of ethics against the coercive powers of the modern state is explored by Jack S. Boozer in his article, 'Children of Hippocrates: Doctors in Nazi Germany', *The Annals of the American Academy of Political and Social Sciences* 450, July 1980, pp. 83–97.

SELECT BIBLIOGRAPHY

These are among the principal works that were consulted for this study. They are divided into separate sections of books and articles. A still useful list of primary source material may be found in Drabkin, Miriam: 1942, 'A Select Bibliography of Greek and Roman Medicine', *Bulletin of the History of Medicine* 11, 4, 399–408. A comprehensive and partially annotated collection of contemporary books and articles may be found in *The Hastings Center's Bibliography of Ethics, Biomedicine, and Professional Responsibility*, ed. by the Hastings Center, University Publications of America, Frederick, Maryland, 1984.

BOOKS

Ackernecht, Edwin H.: 1955, *A Short History of Medicine*, Ronald Press, New York.

Allbutt, Clifford T.:1921, *Greek Medicine in Rome*, MacMillan Ltd., London.

Aristotle: 1910, *Historia Animalium*, translated by D. W. Thompson, Vol. IV of *The Works of Aristotle*, Edited by J. A. Smith and W. D. Ross, 11 vols, Clarendon Press, Oxford.

Aristotle: 1941, *The Basic Works of Aristotle*, Edited by Richard McKeon, Random House, New York.

Aristotle: 1969, *Politics*, translated by H. Rackham, Loeb Classical Library, Harvard University Press, Cambridge.

Barker, Ernest: 1946, *The Politics of Aristotle*, Oxford University Press, Oxford.

Beauchamp, Tom L. and Perlin, Seymour (eds.): 1978, *Ethical Issues in Death and Dying*, Prentice Hall, New Jersey.

Bourke, Vernon J.: *History of Ethics*, 2 vols, Image Books, New York.

Bowra, C. M.: 1959, *The Greek Experience*, Mentor Books, New York.

Breasted, James H.: 1930, *The Edwin Smith Surgical Papyrus*, 2 vols, University of Chicago Press, Chicago.

Burnet, John: 1980, *Early Greek Philosophy*, Macmillan, New York.

Chadwick, John and Mann, W. N.: 1950, *The Medical Works of Hippocrates*, Blackwell Scientific Publications, Oxford.

Choron, Jacques: 1963, *Death and Western Thought*, Macmillan, New York.

Choron, Jacques: 1972, *Suicide*, Charles Scribner's Sons, New York.

Cohn-Haft, Louis: 1956, *The Public Physicians of Ancient Greece*, Smith College Studies in History, Vol. XLII, Smith College, Northampton, Massachusetts.

The Complete Greek Drama: 1938, Edited by Whitney J. Oates and Eugene O'Neill, Jr., 2 vols, Random House, New York.

226   SELECT BIBLIOGRAPHY

Copleston, Frederick: 1962, *A History of Philosophy*, Vol. I, Pt. 2: *Greece and Rome*; Image Books, New York.
Cornford, Francis M.: 1937, *Plato's Cosmology: The Timaeus of Plato*, Bobbs-Merrill, New York.
Cornford, Francis M.: 1957, *From Religion to Philosophy*, Harper, New York.
de Coulanges, Numa D. Fustel: 1956, *The Ancient City: A Study of the Religion, Laws, and Institutions of Greece and Rome*, Doubleday, Garden City, New York.
DeVogel, C. J.: 1966, *Pythagoras and Early Pythagoreanism*, Royal Van Gorcum Ltd., The Netherlands.
Diels, Hermann: 1903, *Die Fragmente der Vorsokratiker*, Weidmann, Berlin.
Diels, Hermann: 1905, *Die Handschriften der antiken Ärzte*, Vol. I: *Hippokrates und Galenos*, n.p., Berlin.
The Dinkard: The Original Pahlavi Text, translated by Ratanshah Erachshah Kohiyar, edited by Peshotan dastur Behramjee Sanjana, Duftur Ashkara Press, Bombay, 1874–1891.
Dodds, E. R.: 1968, *The Greeks and the Irrational*, University of California Press, Berkeley.
Dover, K. J.: 1974, *Greek Popular Morality: In the Time of Plato and Aristotle*, University of California Press, Los Angeles.
Durant, Will: 1939, *The Story of Civilization*, Vol. II: *The Life of Greece*, Simon and Schuster, New York.
Durant, Will: 1944, *The Story of Civilization*, Vol. III: *Caesar and Christ*, Simon and Schuster, New York.
Edelstein, Ludwig: 1931, *Peri Aerōn und die Sammlung der hippokratischen Schriften*, Weidmann, Berlin.
Edelstein, Ludwig: 1967, *Ancient Medicine: Selected Papers of Ludwig Edelstein*, edited by Owsei Temkin and Lilian C. Temkin, Johns Hopkins University Press, Baltimore.
Edelstein, Ludwig and Edelstein, Emma J.: 1945, *Asclepius: A Collection and Interpretation of the Testimonies*, 2 vols., Johns Hopkins University Press, Baltimore.
Entralgo-Laín, P.: 1969, *Doctor and Patient*, translated by F. Partridge, McGraw Hill, New York.
Frankena, William K.: 1963, *Ethics*, Foundations of Philosophy Series, Prentice-Hall, Englewood Cliffs, New Jersey.
Freeman, Kathleen: 1948, *Ancilla to the Pre-Socratic Philosophers: A Complete Translation of the Fragments in Diels, Fragmente der Vorsokratiker*, Basil Blackwell, Oxford.
Galen: 1916, *On the Natural Faculties*, translated by Arthur J. Brock, Loeb Classical Library, Harvard University Press, Cambridge.
Gardner, Helen: 1975, *Art Through the Ages*, edited by Horst de la Croix and Richard G. Tansey, 6th ed., Harcourt Brace Jovanovich, Inc., New York.
Gordon, Benjamin Lee: 1949, *Medicine Throughout Antiquity*, F. A. Davis, Philadelphia.
The Greek Historians: The Complete and Unabridged Works of Herodotus, Thucydides, Xenophon, and Arrian, edited by Francis R. B. Godolphin, 2 vols., Random House, New York, 1942.
Hamilton, Edith: 1969, *Mythology*, New American Library, New York.
Hart, H. L. A.: 1968, *Punishment and Responsibility*, Oxford University Press, Oxford.
Hartman, Edwin: 1977, *Aristotelian Investigations: Substance, Body, and Soul*, Princeton University Press, Princeton.

Heidel, W. A.: 1941, *Hippocratic Medicine: Its Spirit and Method*, Columbia University Press, New York.

Hesiod: *The Works and Days, Theogony, The Shield of Herakles*, translated by Richard Lattimore, University of Michigan Press, Ann Arbor, 1965.

Hicks, R. D.: 1962, *Stoic and Epicurean*, Russell and Russell, New York.

Himes, Norman E.: 1936, *Medical History of Contraception*, Williams and Wilkins Co., Baltimore.

Hippocrates: 1939–1961, *Oeuvres completes d'Hippocrate*, translated by Emile Littré, 10 vols., Javal et Bourdeaux, Paris.

Hippocrates: 1952, *On Intercourse and Pregnancy*, translated by Tage Ellinger, Henry Schuman, Inc., New York.

Hippocrates: 1962, *Medical Works*, edited and translated by W. H. S. Jones, 4 vols., Harvard University Press, Cambridge.

Homer: 1961, *The Iliad of Homer*, translated by Richard Lattimore, University of Chicago Press, Chicago.

Homer: 1967, *The Odyssey of Homer*, translated by Richard Lattimore, Harper and Row, New York.

Jacobovits, Immanuel: 1975, *Jewish Medical Ethics: A Comparative and Historical Study of the Jewish Religion's Attitude and Its Practice*, Bloch Publishing Co., New York.

Jones, W. H. S.: 1924, *The Doctor's Oath*, Cambridge University Press, Cambridge.

Kahn, Charles: 1960, *Anaximander and the Origins of Greek Cosmology*, Columbia University Press, New York.

Kristeller, Paul O.: 1961, *Renaissance Thought: The Classic, Scholastic and Humanistic Strains*, revised edition, Harper and Row, New York.

Kudlien, Fridolf: 1968, *Die Sklaven in der griechischen Medizin der klassischen und hellenistischen Zeit*, Forschungen zur antiken Sklaverei, II, Steiner, Wiesbaden.

Kurtz, Donna and Broadman, John: 1971, *Greek Burial Customs*, Cornell University Press, Ithaca, New York.

Laertius, Diogenes: 1959, *Lives of Eminent Philosophers*, 2 vols., Loeb Classical Library, Harvard University Press, Cambridge.

Largus, Scribonius: 1887, *Compositiones*, edited by Georg Helmreich, Teubner, Leipzig.

Lecky, William E.: 1905, *History of European Morals: From Augustus to Charlemagne*, 2 vols., Longmans, Green, and Co., London.

Leiser, Burton M.: 1973, *Liberty, Justice, and Morals: Contemporary Value Conflicts*, Macmillan, New York.

Lewin, Louis: 1920, *Die Gifte in der Weltgeschichte*, J. Springer, Berlin.

Liddell, H. G. and Scott, R.: 1972, *Greek–English Lexicon (Abridged)*, Clarendon Press, Oxford.

Lucretius: 1913, *On the Nature of Things*, translated by H. A. I. Munro, Bell, London.

Lyons, Albert S. and Petrucelli, Joseph R.: 1978, *Medicine: An Illustrated History*, Harry N. Abrams, Inc., New York.

MacDowell, D. M.: 1963, *Athenian Homicide Law in the Age of the Orators*, University of Manchester Press, Manchester.

MacIntyre, Alasdair: 1976, *A Short History of Ethics*, Mamillan, New York.

Margolis, Joseph: 1975, *Negativities: The Limits of Life*, Charles E. Merrill Publishing Co., Columbus, Ohio.

Michell, Humfrey: 1963, *The Economics of Ancient Greece*, 2nd ed., W. Heffer, Cambridge.

Moon, R. O.: 1909, *The Relation of Medicine to Philosophy*, Longmans, Green, and Co., New York.

Needham, Joseph: 1959, *A History of Embryology*, 2nd ed., revised, Abelard-Schuman, New York.

Neuburger, M.: 1910, *The History of Medicine*, Oxford University Press, London.

Nilsson, Martin: 1964, *A History of the Greek Religion*, translated by F. J. Fielden, 2nd ed., Norton, New York.

Oates, Whitney J. (ed.): 1940, *The Stoic and Epicurean Philosophers*, Random House, New York.

Pauly, August: 1893–1974, *Paulys Real-Encyclopädie der classischen Altertumswissenschaft*, 82 vols., J. B. Metzler, Stuttgart.

Peters, F. E.: 1967, *Greek Philosophical Terms: A Historical Lexicon*, New York University Press, New York.

Plato: 1966, *The Collected Dialogues of Plato*, edited by Huntington Cairns and Edith Hamilton, Random House, New York.

Plato: 1973, *The Republic*, translated by Francis M. Cornford, Oxford Press, London.

Plato: 1976, *The Laws*, translated by Trevor J. Saunders, Penguin Books, New York.

Regan, Tom (ed.): 1980, *Matters of Life and Death*, Random House, New York.

Robinson, John Mansley: 1968, *An Introduction to Early Greek Philosophy*, Houghton Mifflin, New York.

Rohde, Edwin: 1925, *Psyche*, translated by W. B. Hillis, 8th ed., Paul Treuch, Trubner, Ltd., London.

Rosen, George: 1968, *Madness in Society: Chapters in the Historical Sociology of Mental Illness*, University of Chicago Press, Chicago.

Ross, David: 1974, *Aristotle*, Harper and Row, New York.

Ross, W. D.: 1930, *The Right and the Good*, Clarendon Press, Oxford.

Sandbach, F. H.: 1975, *The Stoics*, W. W. Norton, New York.

Saunders, J. B.: 1963, *The Transitions From Ancient Egyptian to Greek Medicine*, University of Kansas Press, Lawrence.

Seneca: 1932, *Moral Essays*, translated by John W. Basore, 3 vols., Loeb Classical Library, William Heinemann, Ltd., London.

Seneca: 1962, *Ad Lucilium Epistulae Morales*, translated by Richard M. Gummere, 3 vols., Harvard University Press, Loeb Classical Library, Cambridge.

Siegel, Rudolph E.: 1968, *Galen's System of Physiology and Medicine*, Karger Press, New York.

Sigerist, Henry E.: 1951 and 1961, *A History of Medicine*, 2 vols., Oxford University Press, New York.

Singer, Charles: 1922, *Greek Biology and Greek Medicine*, Oxford University Press, London.

Singer, Charles and Underwood, E. Ashworth: 1952, *A Short History of Medicine*, 2nd ed., University Press, Oxford.

Snell, Bruno: 1960, *The Discovery of the Mind: The Greek Origins of European Thought*, translated by T. G. Rosenmeyer, Harper and Row, New York.

Sophocles: 1958, *The Oedipus Plays of Sophocles*, translated by Paul Roche, Mentor, New York.

Soranus: 1956, *Soranus' Gynecology*, translated by Owsei Temkin, Johns Hopkins University Press, Baltimore.

Taylor, A. E.: 1928, *A Commentary on Plato's Timaeus*, Oxford University Press, London.

Temkin, Owsei: 1973, *Galenism: The Rise and Decline of a Medical Philosophy*, Cornell University Press, Ithaca.

Theophrastus: 1961, *Inquiry into Plants*, translated by Arthur Hort, 2 vols., Loeb Classical Library, Harvard University Press, Cambridge.

Vaux, Kenneth: 1974, *Biomedical Ethics: Morality for the New Medicine*, Harper and Row, New York.

Vermeule, Emily: 1974, *Aspects of Death in Early Greek Art and Poetry*, The Sather Classical Lectures, Vol. XLVI, University of California Press, Los Angeles, California.

Westermarck, Eduard A.: 1906–08, *The Origins and Development of the Moral Ideas*, 2 vols., Macmillan and Co., London.

Zeller, Eduard: 1962, *The Stoics, Epicureans, and Sceptics*, Russell and Russell, Inc., New York.

The Zend-Avesta: 1880–1883, translated by James Darmester, 3 vols., Clarendon Press, Oxford. Reprinted by Greenwood Press, Westport, Connecticut, 1972.

Zimmern, Alfred E.: 1915, *The Greek Commonwealth*, 2nd ed., rev., Clarendon Press, Oxford.

ARTICLES

Altman, Lawrence K.: 1979, 'The Doctor's World', *New York Times*, May 1, Cl.

Amundsen, Darrel W.: 1973, 'The Liability of the Physician in Roman Law', *International Symposium on Society, Medicine and Law*, edited by H. Karplus, Elsevier, Amsterdam.

Amundsen, Darrel W.: 1974, 'Romanticizing the Ancient Medical Profession: The Characterization of the Physician in the Graeco-Roman Novel', *Bulletin of the History of Medicine* 48, 320–37.

Amundsen, Darrel W.: 1977, 'The Liability of the Physician in Classical Greek Legal Theory and Practice', *Journal of the History of Medicine and Allied Sciences* 32, 172–203.

Amundsen, Darrel W.: 1978, 'History of Medical Ethics: Ancient Near East', *Encyclopedia of Bioethics*, edited by Warren T. Reich, 4 vols., The Free Press, New York, Vol. II, pp. 880–883.

Amundsen, Darrel W.: 1978, 'History of Medical Ethics: Ancient Greece and Rome', *Encyclopedia of Bioethics*, edited by Warren T. Reich, 4 vols., The Free Press, New York, Vol. III, pp. 930–937.

Amundsen, Darrel W.: 1978, 'The Physician's Obligation to Prolong Life: A Medical Duty Without Classical Roots', *Hastings Center Report* 8, 23–30.

Biggarts, J. H.: 1971, 'Cnidus v. Cos', *Ulster Medical Journal* 41, 1–9.

Biggs, Robert: 1969, 'Medicine in Ancient Mesopotamia', *History of Science* 8, 94–105.

Bok, Sissela: 1978, 'Death and Dying: Euthanasia and Sustaining Life', *Encyclopedia of Bioethics*, edited by Warren T. Reich, 4 vols., The Free Press, New York, Vol. I, pp. 268–277.

Boozer, Jack S.: 1980, 'Children of Hippocrates: Doctors in Nazi Germany', *The Annals of the American Academy of Political and Social Sciences: Reflections on the Holocaust* **450**, 83–97.

'Caesarean Section', *Encyclopaedia Britannica*, 1971, Vol. IV.

Clouser, K. Danner: 1977, 'Biomedical Ethics: Some Reflections and Exhortations', *The Monist* **60**, 47–61.

Crawley, A. E.: 1966, 'Foeticide', *Encyclopedia of Religion and Ethics*, edited by James Hastings, 13 vols., Charles Scribner's Sons, New York, Vol. VI, pp. 54–57.

Daube, David: 1972, 'The Linguistics of Suicide', *Philosophy and Public Affairs* **1**, 387–437.

DeLacy, P. H.: 1967, 'Epicurus', *Encyclopedia of Philosophy*, edited by Paul Edwards, 8 vols., Collier Press, New York, Vol. III, pp. 3–5.

Diller, H.: 1933, 'Zur Hippokratesauffassung des Galen', *Hermes* **68**, 167–81.

Edelstein, Ludwig: 1970, 'Hippocrates', *The Oxford Classical Dictionary*, edited by N. G. L. Hammond and H. H. Scullard, 2nd ed., Clarendon Press, Oxford, pp. 518–519.

Ferngren, Gary: 1982, 'A Roman Declamation on Vivisection', *Transactions and Studies of the College of Physicians*, N.S. **IV**, 272–290.

Flacelière, Robert: 1967, 'Medicine', *The Praeger Encyclopedia of Ancient Greek Civilization*, Frederick A. Praeger Pub., New York, pp. 288–290.

Flew, Anthony: 1967, 'Immortality', *Encyclopedia of Philosophy*, edited by Paul Edwards, 8 vols., Collier Press, New York, Vol. IV, pp. 139–150.

Frankena, William K.: 1975, 'The Ethics of Respect for Life', *Respect For Life: In Medicine, Philosophy, and the Law*, edited by Stephen Barker, Johns Hopkins University Press, Baltimore.

Fye, Bruce W.: 1978, 'Active Euthanasia: An Historical Survey of its Conceptual Origins and Introduction into Medical Thought', *Bulletin of the History of Medicine* **62**, 492–502.

Gourevitch, Danielle: 1969, 'Suicide Among the Sick in Classical Antiquity', *Bulletin of the History of Medicine* **43**, 501–18.

Gracia, Diego: 1978, 'The Structure of Medical Knowledge in Aristotle's Philosophy', *Sudhoffs Archiv* **62**, No. 1, 1–36.

Hähnel, R.: 1936, 'Der künstliche Abortus in Altertum', *Archiv für Geschichte der Medizin* **29**, 224–55.

Hall, T. S.: 1974, 'Idiosyncrasy: Greek Medical Ideas of Uniqueness', *Sudhoffs Archiv* **58**, No. 3, 283–302.

Hallie, Philip: 1967, 'Stoicism', *Encyclopedia of Philosophy*, edited by Paul Edwards, 8 vols., Collier Press, New York, Vol. VIII, pp. 19–22.

Hellegers, André E.: 1978, 'Abortion: Medical Aspects', *Encyclopedia of Bioethics*, edited by Warren T. Reich, 4 vols., The Free Press, New York, Vol. I, pp. 1–5.

Hirzel, Rudolf: 1908, 'Der Selbstmord', *Archiv für Religionswissenschaft* **11**, 75–104.

Jaeger, Werner: 1957, 'Aristotle's Use of Medicine as a Model of Method in His Ethics', *Journal of the History of Science* **77**, 54–61.

Kalish, Richard A.: 1978, 'Death, Attitudes Toward', *Encyclopedia of Bioethics*, edited by Warren T. Reich, 4 vols., The Free Press, New York, Vol. I, pp. 286–290.

Kaufman, M. R.: 1966, 'Early Greek Concepts of Mind and "Insanity"', *Psychiatric Quarterly* **40**, 1–33.

Kibre, Pearl: 1945, 'Hippocratic Writings in the Middle Ages', *Bulletin of the History of Medicine* 18, No. 4, 371–412.

Konold, Donald: 1978, 'Codes of Medical Ethics: History', *Encyclopedia of Bioethics*, edited by Warren T. Reich, 4 vols., The Free Press, New York, Vol. I, pp. 162–170.

Kudlien, Fridolf: 1970, 'Medical Ethics and Popular Ethics in Greece and Rome', *Clio Medica* 5, 91–121.

Kudlien, Fridolf: 1973, 'The Old Greek Concept of "Relative Health" ', *Journal of the History of Behavioral Sciences* 9, 53–59.

Kudlien, Fridolf: 1976, 'Medicine as a "Liberal Art" and the Question of the Physician's Income', *Journal of the History of Medicine* 31, 448–59.

Langer, William L.: 1974, 'Infanticide: A Historical Survey', *History of Childhood Quarterly: The Journal of Psychohistory* 1, 353–65.

Longrigg, James: 1963, 'Philosophy and Medicine: Some Early Interactions', *Harvard Studies in Classical Philology* 67, 147–75.

MacKinney, Loren C.: 1952, 'Medical Ethics and Etiquette in the Early Middle Ages: The Persistence of Hippocratic Ideals', *Bulletin of the History of Medicine* 26, 1–31.

Mair, A. W.: 1966, 'Life and Death: Greek', *Encyclopedia of Religion and Ethics*, edited by James Hastings, 13 vols., Charles Scribner's Sons, New York, Vol. VIII, pp. 25–31.

Mair, A. W.: 1966, 'Suicide', *Encyclopedia of Religion and Ethics*, edited by James Hastings, 13 vols., Charles Scribner's Sons, New York, Vol. XII, pp. 26–33.

Miller, Fred: 1976, 'Epicurus on the Art of Dying', *The Southern Journal of Philosophy* 14, No. 2, 169–77.

Moissides, M.: 1932, 'Le Malthusianisme dans l'Antiquite Greque', *Janus* 36, 169–79.

Moravcsik, Julius: 1976, 'Ancient and Modern Conceptions of Health and Medicine', *Philosophy and Medicine* 1, 337–48.

Nittis, Savas: 1939, 'The Hippocratic Oath in Reference to Lithotomy: A New Interpretation with Historical Notes on Castration', *Bulletin of the History of Medicine* 7, 719–28.

Nittis, Savas: 1940, 'The Authorship and Probable Date of the Hippocratic Oath', *Bulletin of the History of Medicine* 8, 1012–21.

Noonan, John T.: 1970, 'An Almost Absolute Value in History', *The Morality of Abortion*, edited by John T. Noonan, Harvard University Press, Cambridge.

Oliver, James H.: 1939, 'An Ancient Poem on the Duties of a Physician', *Bulletin of the History of Medicine* 7, 315–23.

Penella, R. J. and Hall, T. S.: 1973, 'Galen's "On the Best Constitution of our Body:" Introduction, Translation, and Notes', *Bulletin of the History of Medicine* 47, 282–96.

Rose, Herbert J.: 1970, 'Dead, Disposal of the', *The Oxford Classical Dictionary*, edited by N. G. L. Hammond and H. H. Scullard, 2nd ed., Clarendon Press, Oxford, pp. 314–315.

Rosen, George: 1971, 'History in the Study of Suicide', *Psychological Medicine* 1, 267–85.

Rosner, Fred and Mutner, Sussman: 1965, 'The Oath of Asaph', *Annals of Internal Medicine* 63, 317–20.

Sandulescu, C.: 1965, 'Primum Non Nocere: Philological Commentaries on a Medical Aphorism', *Acta Antiqua Hungarica* 13, 359–368.

Singer, Charles and Peck, Arthur: 1970, 'Zoology', *The Oxford Classical Dictionary*, edited by N. G. L. Hammond and H. H. Scullard, 2nd ed., Clarendon Press, Oxford, pp. 1148–1150.

Solmsen, Friedrich: 1957, 'The Vital Heat, The In-Born Pneuma and the Aether', *The Journal of Hellenic Studies*, No. 77, 119–23.

Strachan, J. C. G.: 1970, 'Who *Did* Forbid Suicide at *Phaedo* 62 B?', *Classical Quarterly* 20, 216–220.

Temkin, Owsei: 1953, 'Greek Medicine as Science and Craft', *Isis* 33, 213–25.

Temkin, Owsei: 1973, 'Health and Disease', *Dictionary of the History of Ideas*, edited by Philip P. Wiener, 4 vols., Charles Scribner's Sons, New York, Vol. II, pp. 395–407.

Temkin, Owsei: 1975, 'The Idea of Respect for Life in the History of Medicine', *Respect For Life: In Medicine, Philosophy, and the Law*, edited by Stephen Barker, Johns Hopkins University Press, Baltimore.

Veatch, Robert M.: 1978, 'Codes of Medical Ethics: Ethical Analysis', *Encyclopedia of Bioethics*, edited by Warren T. Reich, 4 vols., The Free Press, New York, Vol. I, pp. 172–179.

Vlastos, Gregory: 1953, 'Isonomia', *American Journal of Philology* 74, No. 4, 337–66.

Warren, Mary Anne: 1973, 'On the Moral and Legal Status of Abortion', *The Monist* 57, 43–61.

Wilson, John A.: 1962, 'Medicine in Ancient Egypt', *Bulletin of the History of Medicine* 36, 114–23.

INDEX

abortion: general references to, 81–87, 99–125 *passim*, 161–169, 182, 209 n, 210 n., 222 n.; and contraception, 104–107; and doing no harm, 156–158; Egyptian views of, 63; methods of, 84–86, 107–108, 152, 206 n., 3 207 n.; and midwives, 109; Near Eastern views of, 63; *Roe vs Wade*, 175; Aristotle on, 82, 115–119, 124–125, 213 n.; Edelstein on, 84–86; Plato on, 82, 112–115, 124–125, 212 n.; Pythagoreans on, 81–87 *passim*, 110–112, 124–125; Seneca on, 119–125, 214 n., 215n.; Soranus on, 85, 105–106, 152–153, 206 n.; Stoics on, 119–123, 173, 215 n. (*see also* duty, Hippocratic Oath)

Achilles, 41

Ackerknecht, Erwin, 195 n.

adultery, 78, 79–80, 102, 114, 205 n., 210 n., 212 n.

Aelian, 208 n.

Aeschylus, 46, 47, 201 n.

Aexis, 205 n.

Alcmaeon, 17, 33

Alexandria: library at, 65–66

Allbutt, T. Clifford, 7, 196 n.

Altman, Lawrence, 223 n.

Amundsen, Darrel, 59, 171, 196 n., 197 n., 202 n., 204 n., 205 n., 222 n.

Anacreon, 46

anatomy: and warfare, 91; and vivisection, 171

Anaximander, 43, 44, 45, 54

Anaximenes, 43, 44, 45, 55

Antoninus, 210 n.

Apollo, 69, 96, 176

Apuleius, 206 n., 211 n., 221 n.

Aristophanes, 216 n.

Aristotle: general references to, xvii, xxiii, xxiv, 4, 5, 9, 12, 13, 15, 26, 28–33 *passim*, 37, 49–52, 55. 56, 64, 81–82, 99, 104, 110, 115–119, 124–125, 131, 133, 141–144, 148–149, 161, 165, 179, 210 n., 219 n.; on abortion, 82, 115–119, 124–125, 213 n.; on contraception, 104; on courage, 142–143; on death, 50–52, 55, 142–143, 166; on embryology, 117–119, 213 n.; on ethics and medicine, 11; on euthanasia, 141–144 *passim*; on the family, 115–116; on the four causes, 29; on health, 28–30, 116, 119 n.; and Hippocrates, 28; on infanticide, 115–119 *passim*; on moral responsibility, 141–142; on nature, 28; on old age, 29–30; on over-population, 115–118; on physicians, 4–5, 11–13; and respect for human life, 143–144; on the soul, 50–52, 116–119, 213 n.; on suicide, 81–82, 141–144, 218 n.; on virtue, 142–143; *De Anima*, 50–51, 201 n., 213 n.; *De Caelo*, 202 n.; *De Generatione Animalium*, 199 n.; *De Partibus Animalium*, 199 n.; *Historia Animalium*, 118, 210 n., 213 n.; *Metaphysics*, 199 n.; *Nichomachean Ethics*, 11, 51, 141–142, 197 n., 201 n., 213 n., 216 n., 218 n.; *Physics*, 28, 199 n.; *Politics*, 9, 115–116, 124, 195 n., 196 n., 197 n., 202 n., 203 n., 212 n., 213 n.; *Rhetoric*, 51, 201 n., 213 n.

Aristoxenus, 74, 75, 76, 204 n.

233

PALLAS PAPERBACKS